SpringerBriefs in Plant Science

For further volumes:
http://www.springer.com/series/10080

Yuyang Zhang

Ascorbic Acid in Plants

Biosynthesis, Regulation and Enhancement

 Springer

Yuyang Zhang
Key Laboratory of Horticultural Plant Biology
Ministry of Education, College of Horticulture
  and Forestry Sciences
Huazhong Agricultural University
Wuhan, Hubei
China

ISSN 2192-1229   ISSN 2192-1210   (electronic)
ISBN 978-1-4614-4126-7   ISBN 978-1-4614-4127-4   (eBook)
DOI 10.1007/978-1-4614-4127-4
Springer New York Heidelberg Dordrecht London

Library of Congress Control Number: 2012942715

Printed on acid-free paper

Springer is part of Springer Science+Business Media (www.springer.com)

Preface

Ascorbic acid (vitamin C) is synthesized from hexose sugars. Ascorbic acid is an important antioxidant and redox buffer in plants, playing important roles in metabolism and plant responses to abiotic stresses and pathogens. It also works as an enzyme cofactor, so it has multiple roles in various plant physiological processes. Humans have lost the ability to synthesize ascorbate and have to absorb ascorbic acid from the diet including fresh fruits and vegetables, as they are the major sources of ascorbate.

Several pathways for ascorbic acid biosynthesis and metabolism have been identified in plants since 1998. Significant progresses have been made in relation to key enzymes and genes involved in the ascorbate biosynthesis and metabolism. Biochemical and molecular genetic evidence supports synthesis from GDP-D-mannose via L-galactose (D-Man/L-Gal pathway) as a significant source of ascorbic acid. More recently, evidence for pathways via uronic acids has been obtained: overexpression of myoinositol oxygenase, D-galacturonate reductase and L-gulono-1,4-lactone oxidase all increase ascorbic acid concentration in plants.

An understanding of how ascorbate is synthesized should provide a basis for engineering or otherwise manipulating its accumulation. However, in the examples of pathway engineering so far, the increase in ascorbic acid has been modest on an absolute or proportional basis. Therefore, a deeper understanding of ascorbic acid metabolism is needed to achieve larger increases. Identifying genes that control ascorbate accumulation may hold promise, particularly if regulatory genes can be identified. Recently, more attention has been paid to the control and regulation of ascorbic acid biosynthesis, as it is constantly regulated by the plant development and the environmental factors, e.g., light. Ascorbic acid is also frequently reported to affect plant growth and development, e.g., flowering time and fruit ripening. The increasing knowledge about ascorbic acid regulation should facilitate engineering or modulating its accumulation. Besides the metabolic engineering from genes, the environmental regulation may stand close to industrial practice.

The author takes this opportunity to give a brief review covering the biological function, biosynthesis and metabolism, regulation, and metabolic modification of ascorbate in plants. Processing and storage of plant product is not included, though it decides the final amount of this nutrient. The outstanding work on ascorbic acid in plants by scientific colleagues is greatly acknowledged.

Acknowledgments

The author acknowledges the excellent work by scientists working on ascorbate in plants, while many of them are not cited because of space limitation. The author thanks Professor Zhibiao Ye, Professor Hanxia Li, and Professor Xiuxin Deng for critical comments on the manuscript outline and earlier draft. Part of the ascorbate research work on tomato described here was carried out in Professor Zhibiao Ye's laboratory. The author wishes to express his sincere thanks to Dr. Chanjuan Zhang and Dr. Liping Zou for providing valuable discussion and information. Huge thanks are owed to Hannah Smith for her continuous encouragement and professional assistance. Part of the author's work is supported by the National Program on Key Basic Research Project (2011CB100600) and Natural Science Foundation of China (31171974).

Acknowledgments

Contents

1 Discovery and Determination of Ascorbic Acid 1
  1.1   Chemical Nature. 1
  1.2   Determination of Ascorbate in Plants 3
  References 5

2 Biological Role of Ascorbate in Plants 7
  2.1   Role in Plant Growth and Development 8
        2.1.1   Cell Division 9
        2.1.2   Cell Wall Metabolism and Cell Expansion 9
        2.1.3   Shoot Apical Meristem Formation 11
        2.1.4   Root Development 16
        2.1.5   Photosynthesis 16
        2.1.6   Regulation of Florescence 17
        2.1.7   Regulation of Leaf Senescence 18
  2.2   The Cofactors for Enzyme Activity 20
  2.3   Plant Antioxidation Capacity 21
  2.4   Heavy Metal Evacuation and Detoxification 23
  2.5   Role in Stress Defense 23
  References 28

3 Ascorbate Biosynthesis in Plants 33
  3.1   D-Man/L-Gal Pathway 34
  3.2   Myoinositol Pathway 35
  3.3   Galacturonate Pathway 36
  3.4   Gulose Pathway 38
  3.5   VTC2 Cycle 38
  References 40

4   The Oxidization and Catabolism of Ascorbate   43
       References   45

5   Recycling of Ascorbate   47
       References   47

6   Distribution and Transport of Ascorbate   49
       References   52

**7   Enzymes Involved in Ascorbate Biosynthesis
       and Metabolism in Plants**   55
       7.1   Phosphomannose Isomerase (PMI: EC 5.3.1.8).   55
       7.2   Phosphomannomutase (PMM: EC 5.4.2.8).   56
       7.3   GDP-D-Mannose Pyrophosphorylase (GMP: EC 2.7.7.22)   57
       7.4   GDP-D-Mannose-3,5-Epimerase (GME: EC 5.1.3.18)   59
       7.5   GDP-L-Galactose Phosphorylase (GGP: EC 2.7.7.69)   61
       7.6   L-Galactose-1-Phosphate Phosphatase (GPP: EC 3.1.3.25)   62
       7.7   L-Galactose Dehydrogenase (GalDH: EC 1.1.1.117)   64
       7.8   L-Galactono-1,4-Lactone Dehydrogenase
              (GLDH: EC 1.3.2.3)   65
       7.9   Ascorbate Oxidase (AO: EC 1.10.3.3).   66
       7.10   Ascorbate Peroxidase (APX: EC 1.11.1.11)   68
       7.11   Dehydroascorbate Reductase (DHAR: EC.1.8.5.1)   71
       7.12   Monodehydroascorbate Reductase (MDHAR: EC.1.6.5.4)   72
       7.13   Myoinositol Oxygenase (MIOX: EC 1.13.99.1)   73
       7.14   Alternative Pathway and Enzyme for Ascorbate
              Biosynthesis and Metabolism   74
              7.14.1 L-Gulono-1,4-Lactone Oxidase
                      (GLOase: EC 1.1.3.8)   74
              7.14.2 D-Galacturonate Reductase (GalUR: EC 1.1.1.203). . . .   75
       7.15   The Multi-Gene Family Involved in Ascorbate
              Biosynthesis and Metabolism   75
       References   77

8   Regulation of Ascorbate Synthesis in Plants.   85
       8.1   Growth and Postharvest Condition   85
       8.2   Light Regulation.   87
       8.3   Plant Growth Regulator.   89
       8.4   Jasmonates Regulation   90
       8.5   Feedback Regulation.   90
       8.6   Coordination and Compensation.   91
       8.7   Transcription Factor   93
       References   94

9   Ascorbate in Tomato, a Model Fruit   99
     References   103

10   Metabolic Modification of Ascorbate in Plants   105
     10.1   Overexpression and Ectopic Expression   106
     10.2   Gene Suppression   108
     References   109

**11   Regulating Ascorbate Biosynthesis and Metabolism
     for Stress Tolerance in Plants**   111
     References   114

Abbreviations

ABA	Abscisic acid
ACC	1-Aminocyclopropane-1-carboxylate
AO	Ascorbate oxidase
APX	Ascorbate peroxidase
CAT	Catalase
cDNA	Complementary DNA
DHAR	Dehydroascorbate reductase
DTT	Dithiothreitol
FAD	Flavin adenine-dinucleotide
GA	Gibberellins
GalDH	L-Galactose dehydrogenase
GalUR	D-Galacturonate reductase
GFP	Green fluorescent protein
GGP	GDP-L-Galactose phosphorylase
GLDH	L-Galactono-1,4-lactone dehydrogenase
GLOase	L-Gulono-1,4-lactone oxidase
GME	GDP-D-mannose-3,5-epimerase
GMP	GDP-D-mannose pyrophosphorylase guanosine monophosphate
GPP	L-galactose-1-phosphate phosphatase
GR	Glutathione reductase
GUS	β-Glucuronidase
HPLC	High performance liquid chromatography
JA	Jasmonic acid
MDHAR	Monodehydroascorbate reductase
MeJA	Methyl jasmonate
MIOX	Myinositol oxygenase
mRNA	Messenger RNA
NADPH	Reduced nicotinamide adenine dinucleotide phosphate
PEG	Polyethylene glycol
PMI	Phosphomannose isomerase
PMM	Phosphomannomutase

POD	Peroxidase
QTL	Quantitative trait loci
RFLP	Restriction fragment length polymorphisms
RNA	Ribonucleic acid
RNAi	RNA interference
ROS	Reactive oxygen species
SA	Salicylic acid
SOD	Superoxide dismutase
UV	Ultraviolet
vtc	Ascorbate-deficient mutant in Arabidopsis

Chapter 1
Discovery and Determination
of Ascorbic Acid

Ascorbic acid (also named vitamin C) has important antioxidant and metabolic
functions in both plants and animals, but humans, and a few other animal species,
have lost the capacity to synthesize it. Plant-derived ascorbic acid is thus the major
source of vitamin C in the human diet. Although the importance of ascorbic acid in
human health has been realized for three centuries, the final identification of this
essential nutrient molecule came at twentieth century after continuous efforts in
medicine, philology and chemistry.

1.1 Chemical Nature

Ascorbic acid is an organic compound with antioxidant properties. The molecular
formula of ascorbic acid is $C_6H_8O_6$ with the molecular weight of 176.13 (Fig. 1.1).
It is a white solid, but impure samples can appear yellowish. It dissolves well in
water to give mildly acidic solutions. Ascorbic acid can be slightly dissolved in
ethanol, but not in diethyl ether, chloroform, benzene, petroleum ether or lipid.

Ascorbic acid can be easily oxidized acting as a strong reducer. It is oxidized
and degraded when heated or in solution, and more unstable under alkalinity
condition. As a reducing agent, ascorbic acid degrades upon exposure to air,
converting the oxygen to water. The redox reaction is accelerated by the presence
of metal ions and light. It can be oxidized by one electron to a radical state or
doubly oxidized to the stable form called dehydroascorbate.

The name of ascorbic acid came from the letter of a-(meaning "no") and the
word of scorbutus (meaning "scurvy"), the disease caused by a deficiency of
vitamin C. Because the ascorbic acid is derived from glucose, many animals are
able to produce it, but humans require it as part of their nutrition. Other vertebrates
lacking the ability to produce ascorbic acid include primates, guinea pigs, teleost
fishes, bats, and birds, all of which require it as a dietary micronutrient.

Y. Zhang, *Ascorbic Acid in Plants*, SpringerBriefs in Plant Science,
DOI: 10.1007/978-1-4614-4127-4_1, © The Author 2013

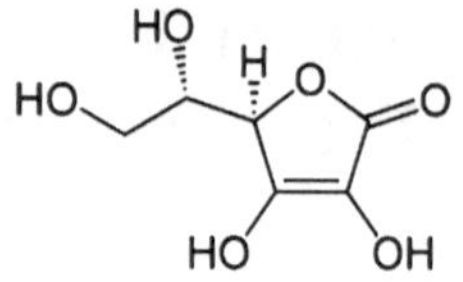

Fig. 1.1 The molecular
structure of ascorbic acid

Thus ascorbic acid is commonly used for deficiency-related disease such as scurvy and helps improve immunity.

From the middle of the eighteenth century, it was noted that lemon juice could prevent sailors from getting scurvy. At first it was supposed that the acid properties were responsible for this benefit; however, it soon became clear that other dietary acids, such as vinegar, had no benefits. In 1907, two Norwegian physicians reported an essential disease-preventing compound in foods that was distinct from the one that prevented beriberi. These physicians were investigating dietary deficiency diseases using the new animal model of guinea pigs, which are susceptible to scurvy. The newly discovered food-factor was eventually called vitamin C.

From 1928 to 1932, the Hungarian research team led by Albert Szent-Györgyi, as well as that of the American worker Charles Glen King, identified the antiscorbutic factor as a particular single chemical substance. At the Mayo clinic, Szent-Györgyi had isolated the chemical hexuronic acid from animal adrenal glands. He suspected it to be the antiscorbutic factor, but could not prove it without a biological assay. This assay was finally conducted at the University of Pittsburgh in the laboratory of King, which had been working on the problem for years, using guinea pigs. In late 1931, King's lab obtained adrenal hexuronic acid indirectly from Szent-Györgyi and using their animal model, proved that it is vitamin C, by early 1932.

This was the last of the compound from animal sources, but, later that year, Szent-Györgyi's group discovered that paprika pepper, a common spice in the Hungarian diet, is a rich source of hexuronic acid. He sent some of the isolated chemical to Walter Norman Haworth, a British sugar chemist. In 1933, working with the then-Assistant Director of Research Edmund Hirst and their research teams, Haworth deduced the correct structure and optical-isomeric nature of vitamin C, and in 1934 reported the first synthesis of the vitamin. In honor of the compound's antiscorbutic properties, Haworth and Szent-Györgyi now proposed the new name of "a-scorbic acid" for the compound. It was named L-ascorbic acid by Haworth and Szent-Györgyi when its structure was finally proven by synthesis. In 1937, the Nobel Prize for chemistry was awarded to Norman Haworth for his work in determining the structure of ascorbic acid (shared with Paul Karrer, who received his award for work on vitamins), and the prize for Physiology or Medicine that year went to Albert Szent-Györgyi for his studies of the biological functions of L-ascorbic acid.

Because ascorbic acid is acidic, it tends to act as an acid when mixed with water. Ascorbic acid loses one of its hydrogen atoms, resulting in production of a closely related molecule called ascorbate, with the chemical formula $C_6H_7O_6$.

Because they're so closely related, it's not uncommon for scientists to refer to ascorbic acid as ascorbate or vice versa. In latter part of this book, the author refers to ascorbic acid as ascorbate.

Ascorbate usually acts as an antioxidant. It typically reacts with oxidants of the reactive oxygen species (ROS), such as the hydroxyl radical formed from hydrogen peroxide. Such radicals are damaging to animals and plants at the molecular level due to their possible interaction with nucleic acids, proteins, and lipids. Sometimes these radicals initiate chain reactions. Ascorbate can terminate these chain radical reactions by electron transfer. Ascorbic acid is special because it can transfer a single electron, owing to the stability of its own radical ion of dehydroascorbate.

The oxidized forms of ascorbate are relatively unreactive, and do not cause cellular damage. However, being a good electron donor, excess ascorbate in the presence of free metal ions can not only promote but also initiate free radical reactions, thus making it a potentially dangerous pro-oxidative compound in certain metabolic contexts.

1.2 Determination of Ascorbate in Plants

Variety of methods including titration, spectrophotometry, potentiometry, voltametry, coulmetry, and high performance liquid chromatography (HPLC) have been applied for the determination of ascorbate [1].

The traditional way to analyze the ascorbate content is titration with an oxidizing agent, and several procedures have been developed, mainly relying on iodometry. Iodine is used in the presence of a starch indicator. Iodine is reduced by ascorbic acid, and, when all the ascorbic acid has reacted, the iodine is then in excess, forming a blue-black complex with the starch indicator. This indicates the end-point of the titration. Another organic dye, 2,6-dichloroindophenol (DCIP), is also utilized in titration method to determine the ascorbate under the similar mechanism. Also the methods for the determination of ascorbate in plants utilizing a titrimetric method with potassium brommat-bromide solution [2] and N-bromosuccinimide [3] are described.

Colorimetric method is also classical approach for ascorbate assay, with several improved derivatives [4]. The Folin-Ciocalteu method was optimized to determine ascorbate in extract of fruit products [5]. And UV-spectrophotometric method was also utilized for the determination of ascorbate in plants [6, 7].

It was found that an iodine-modified platinum electrode gives a linear potentiometric response to ascorbate, and thus is used for determination of ascorbate [8].

A high-resolution selected ion monitoring gas chromatography–mass spectrometry methodology was developed to analyze ascorbate and dehydroascorbate in plant extracts [9]. The quantitative analysis is based on the *tert.*-butyldimethylsilyl derivatives of ascorbate and dehydroascorbate, with an isotope dilution assay using $[^{13}C_1]$ ascorbate and $[^{13}C_1]$ dehydroascorbate. The reproducibility of the whole quantification procedure was very high: giving a relative standard deviation (R.S.D.) of 4 and 10 % for ascorbate and dehydroascorbate, respectively. The reproducibility of the analysis of the redox ratio of ascorbate/(ascorbate + dehydroascorbate) was even higher (R.S.D. = 1 %) [9].

Ascorbate in plant tissues can be analyzed by high-performance capillary zone electrophoresis. The capillary electrophoresis method for the determination of ascorbate in vegetative tissues has superior resolution comparable to HPLC separations, and a comparable analysis time, but lower sensitivity [10]. The capillary zone electrophoresis procedure for the determination of ascorbate and total ascorbate (ascorbate + dehydroascorbate) in vascular and nonvascular plants is described [11], and is improved for simultaneous determination of ascorbate and glutathione [12]. In addition, capillary electrophoretic method for the determination of ascorbate in a single protoplast, i.e., a plant cell without cell wall, has been developed [13].

Now the ascorbate is normally determined by HPLC with more accuracy and high throughput, and several derivations have been developed and optimized [14–18].

Ascorbate and dehydroascorbate can be simultaneously assayed in plant materials by HPLC [16]. Ascorbate was detected directly at 248 mm whilst dehydroascorbate was simultaneously detected as the 1,2-phenylenediamine derivative at 348 mn. Resolution and accuracy were shown to be satisfactory for naturally occurring amounts in plant materials [16]. A radio-HPLC method was also developed to determine the in vivo ^{14}C-labeled ascorbate and dehydroascorbate in plant cell suspensions upon incubation of cultures with exogenous d-[14C]mannose [17].

The HPLC protocol was optimized to reduce the analysis times to only 6 min, making the method suitable for the high-resolution screening of large numbers of samples. It should also be noted that the inclusion of the reducing agent dithiothreitol as a "stabilizer" in extracts with high phenolic content actually promoted oxidation of these antioxidants [18].

Here, a HPLC detection method for ascorbate in tomato is represented as an example for ascorbate determination in plants. About 0.2–1 g tomato tissue is sampled and quickly frozen in liquid nitrogen and stored in −80 °C. Five milliliter of pre-cold 0.1 % (w/v) HPO_3 is added to sample, and then the sample is subjected to 30 min extraction, and 12,000 g centrifugation at 4 °C for 10 min. The supernatant is filtered thought Millipore membrane (0.22 μm) to a new centrifuge tube. The filtrate is kept in dark and placed on ice. Three hundred microliters of supernatant together with equal volume of 50 mmol/L DTT is mixed and kept under dark for 15 min. And then the reaction mix is used for HPLC analysis to determine the total ascorbate content. The chromatographic column is SB-aq C18 from Agilent. Mobile phase is 0.2 M acetate buffer (pH 4.5). The flow rate is 1.0 ml/min. The detection wavelength is 254 nm. Standard ascorbate solution of 2–40 mg/L is used for HPLC analysis and preparing standard curve (Fig. 1.2).

The preparation of 0.2 mol/L acetate buffer (pH 4.5):

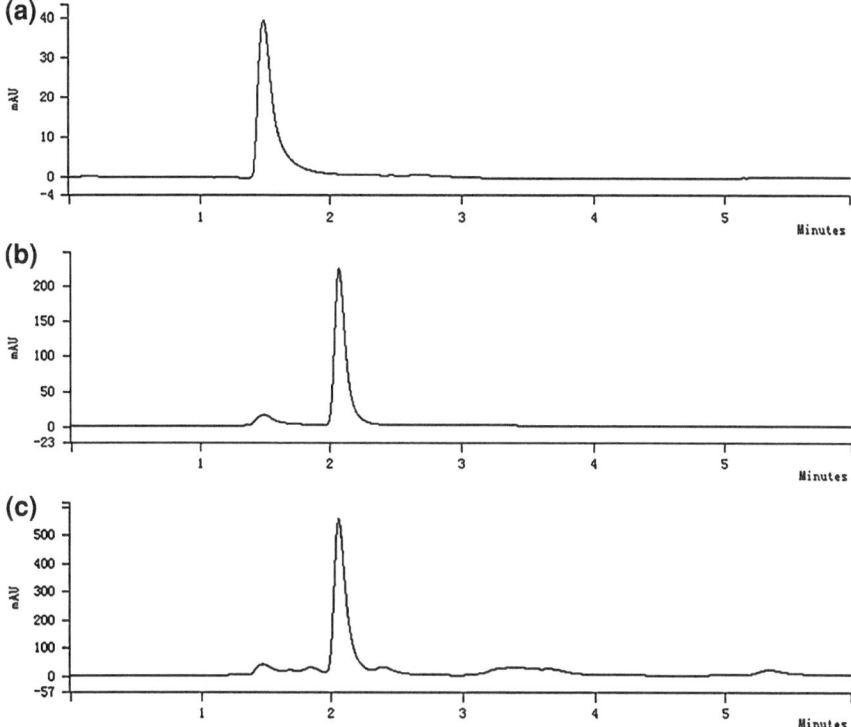

Fig. 1.2 HPLC chromatogram of ascorbate **a** HPO$_3$ solution; **b** ascorbate standard solution; **c** tomato sample

Firstly prepare the 0.2 mol/L NaAc solution (27.22 g NaAc.3H$_2$O per liter) and 0.2 mol/L HAc solution (11.5 mL glacial acetic acid per liter). And then mix 215 mL NaAc solution and 285 mL HAc solution to generate 0.2 mol/L acetate buffers (pH4.5).

The preparation of 50 mmol/L DTT solution:

The DTT powder is dissolved in 0.01 mol/L NaAc solution (pH 5.2) to give the final concentration of 50 mmol/L DTT. The 50 mmol/L DTT solution is vacuum-filtered, aliquoted, and stored at −20 °C.

References

1. Iqbal Y, Ihsanullah I, Shaheen N, Hussain I (2009) Significance of vitamin C in plants. J Chem Soc Pak 31:169–170
2. Birghila S, Dobrinas S, Matei N, Magearu V, Popescu V, Soceanu A (2004) Distribution of Cd, Zn and ascorbic acid in different stages of tomato (*Lycopersicum esculentum Solanaceae*) plant growing. Rev Chim-Bucharest 55:683–685

3. Ogunlesi M, Okiei W, Azeez L, Obakachi V, Osunsanmi M, Nkenchor G (2010) Vitamin C contents of tropical vegetables and foods determined by voltammetric and titrimetric methods and their relevance to the medicinal uses of the plants. Int J Electrochem Sci 5:105–115

4. Gillespie KM, Ainsworth EA (2007) Measurement of reduced, oxidized and total ascorbate content in plants. Nat Protoc 2:871–874

5. George S, Brat P, Alter P, Amiot MJ (2005) Rapid determination of polyphenols and vitamin C in plant-derived products. J Agric Food Chem 53:1370–1373

6. Hussain I, Khan L, Marwat GA, Ahmed N, Saleem M (2008) Comparative study of vitamin C contents in fruits and medicinal plants. J Chem Soc Pak 30:406–409

7. Hussain I, Khan L, Marwat GA (2011) Analysis of vitamin C in selected medicinal plants. J Chem Soc Pak 33:260–262

8. Abdullin IF, Turova EN, Ziyatdinova GK, Budnikov GK (2002) Potentiometric determination of ascorbic acid: Estimation of its contribution to the total antioxidant capacity of plant materials. J Anal Chem 57:353–355

9. Wingsle G, Moritz T (1997) Analysis of ascorbate and dehydroascorbate in plant extracts by high-resolution selected ion monitoring gas chromatography mass spectrometry. J Chromatogr A 782:95–103

10. Davey MW, Bauw G, VanMontagu M (1996) Analysis of ascorbate in plant tissues by high-performance capillary zone electrophoresis. Anal Biochem 239:8–19

11. Herrero-Martinez JM, Simo-Alfonso E, Deltoro VI, Calatayud A, Ramis-Ramos G (1998) Determination of L-ascorbic acid and total ascorbic acid in vascular and nonvascular plants by capillary zone electrophoresis. Anal Biochem 265:275–281

12. Herrero-Martinez JM, Simo-Alfonso EF, Ramis-Ramos G, Deltoro VI, Calatayud A, Barreno E (2000) Simultaneous determination of L-ascorbic acid, glutathione, and their oxidized forms in ozone-exposed vascular plants by capillary zone electrophoresis. Environ Sci Technol 34:1331–1336

13. Olsson J, Nordstrom O, Nordstrom AC, Karlberg B (1998) Determination of ascorbic acid in isolated pea plant cells by capillary electrophoresis and amperometric detection. J Chrom A 826:227–233

14. Rizzolo A, Forni E, Polesello A (1984) HPLC assay of ascorbic-acid in fresh and processed fruit and vegetables. Food Chem 14:189–199

15. Ohta K, Harada K (1996) Studies on the measurement of ascorbic acid and dehydroascorbic acid in tea plants. Nippon Nogeik Kaishi 70:873–882

16. Tausz M, Kranner I, Grill D (1996) Simultaneous determination of ascorbic acid and dehydroascorbic acid in plant materials by high performance liquid chromatography. Phytochem Anal 7:69–72

17. Wolucka BA, Davey MW, Boerjan W (2001) A high-performance liquid chromatography radio method for determination of L-ascorbic acid and guanosine 5′-diphosphate-L-galactose, key metabolites of the plant vitamin C pathway. Anal Biochem 294:161–168

18. Davey MW, Dekempeneer E, Keulemans J (2003) Rocket-powered high-performance liquid chromatographic analysis of plant ascorbate and glutathione. Anal Biochem 316:74–81

Chapter 2
Biological Role of Ascorbate in Plants

Ascorbate, as a strong and active antioxidant, plays an important role in keeping human health. It helps promote iron absorption, increase antibody concentration and improve human immunity. Ascorbate is used to regenerate α-tocopherol (vitamin E), and remove radicals which could induce cancer and senescence. Ascorbate is beneficial to strengthen blood vessel and decrease cholesterol concentration. Ascorbate can help prevent arteriosclerosis related cardiovascular diseases as well as hypertension and apoplexy. Ascorbate prevents disease associated with connective tissue (e.g., scurvy), promotes the formation of collagen and development of cartilage, strengthens tooth, tightens skin and facilitates wound healing [1].

In contrast to most animals, humans lack the ability to synthesize ascorbate as a result of a mutation in the last enzyme required for ascorbate biosynthesis. Human must absorb ascorbate from dietary sources, especially plant products such as fresh fruit and vegetable [2].

Consistent with its multi-function in human and animals, ascorbate in plants has beneficial influences on various aspects in plants. Through modifying gene expression, ascorbate not only act to regulate defense and survival but also act via phytohormones to modulate plant growth [3]. Emerging research results indicate that ascorbate, existing widely in plants as the abundant micromolecule substance, fulfils its essential roles in series of physiological processes such as plant defense against oxidization, co-factor of key enzymes, plant cell division, cell expansion, growth and development, and senescence [4–6]. Ascorbate, at least in some plant species, is also the substrate for the biosynthesis of oxalate and tartrate [7, 8].

However, the multifunction of ascorbate makes it complicated to decipher its exact role under certain physiological process. *Arabidopsis thaliana* mutants lacking ascorbate cannot grow well, but it is not known which function is critical: control of reactive oxygen or the proposed roles in modulating cell expansion and division [9]. Thus, there is need to increase our understanding of this enigmatic molecule since it could be involved in a wide range of important functions from

Y. Zhang, *Ascorbic Acid in Plants*, SpringerBriefs in Plant Science, DOI: 10.1007/978-1-4614-4127-4_2, © The Author 2013

antioxidant defence and photosynthesis to growth regulation. The collection of *vtc* mutants should contribute to a deeper understanding of the proposed functions of this multifaceted molecule.

2.1 Role in Plant Growth and Development

Ascorbate is an important growth regulator as well as antioxidant for plants. Both the ascorbate mutants in Arabidopsis and the various transgenic plant species with relation to ascorbate synthesis are frequently reported to show altered growth and development. Up to date, the plant mutants without ascorbate have not been identified, which may be a demonstration that the ascorbate is an indispensable molecule for plants to survive [10]. The much low amount of ascorbate (15 % of the wild-type control) in transgenic tomato with gene suppression of GDP-D-mannose-3,5-epimerase (GME) leads to severe growth retardation (unpublished data). These evidences together with the fact that ascorbate accumulation level varies with developmental stages in plants strongly indicate that the ascorbate is closely involved in plant growth and development.

Ascorbate as well as its metabolism related enzymes is involved in the control of plant growth processes. Ascorbate modulates plant growth probably by controlling several basic biological processes, such as (i) the biosynthesis of hydroxyproline-rich proteins required for the progression of G1 and G2 phases of the cell cycle, (ii) the crosslinking of cell wall glycoproteins and other polymers, and (iii) redox reactions at the plasmalemma involved in elongation mechanisms. The ascorbate free radical induces a high vacuolization responsible for elongation. This effect may be dependent on the activity of the redox system linked to the plasmalemma. The modulation of the plasmalemma energetic state derived from the ascorbate-induced hyperpolarization and the activity of an intrinsic transplasmalemma ascorbate-regenerating enzyme has been the basis of ascorbate-mediated plant growth regulation.

Recent research results propose that the mechanism for ascorbate regulation of plant growth and development may reside in its interaction with phytohormones. Ascorbate is cofactor for biosynthesis of several phytohormones such as ethylene, gibberellins (GA) and abscisic acid (ABA). The endogenous ascorbate influences the biosynthesis of phytohormones, as well as the signal transduction pathway mediated by phytohormones [3]. The ascorbate in leaves could regulate the plant growth through interaction with phytohormones [3]. Transcript changes indicate that growth and development are constrained in ascorbate defective mutant, *vtc1*, by the modulation of abscisic acid signaling. Abscisic acid contents are significantly higher in *vtc1* than that in the wild-type [3].

The onion root has been the model for investigating the role of ascorbate on plant growth [11]. Exploring how ascorbate participates in the plant growth and development regulation as an antioxidant will deepen insights into the biological function of the ascorbate.

2.1.1 Cell Division

Ascorbate is frequently reported to be related with cell division in plants. The ascorbate content in the meristem is usually higher than that in cell division-inactive tissue, such as quiescent center of the maize root tip. This is consistent with fact that the expression level of ascorbate oxidase gene (AO), which acts in ascorbate oxidation and metabolism, in the quiescent center was higher than that in the surrounding meristem cells [12]. Ascorbate has been implicated in regulation of cell division by influencing progression from G1 to S phase of the cell cycle [4]. The exogenous ascorbate promoted the G1 to S progression in root meristem and pericycle of onion resulting in decreasing cell number in the quiescent center.

The ascrobate and dehydroascorbate exerted significant impact on the process of cell division in the tobacco suspension cells [13]. Lycorine, the inhibitor of ascorbate, could prevent the cell division, while supplementary ascorbate could restore the cell division [14]. The decreasing of endogenous ascorbate resulted in retarded cell division, lowered growth rate of young branches, and slow plant growth, as demonstrated both Arabidopsis mutant [15] and transgenic tobacco [16].

The plasmalemma is a dynamic interface that perceives and transmits information concerning changes in the environment to the nucleus to modify gene expression. In plants, ascorbate is an essential part of this dialogue. The concentration and ratio of reduced to oxidized ascorbate in the apoplast possibly modulates cell division and growth [5]. Although ascorbate consumption is more or less the same during cell division and cell expansion, the ascorbate/monodehydroascorbate ratio is 6–10 during cell division and 1–3 during cell expansion, indicating the reduced state of ascorbate is more required for cell division [14].

Callus formation, the process in which cell division is vigorously activated, is also affected by ascorbate. Foliar application with ascorbate is shown to promote the callus formation from the cut surfaces of scion stems, and eventually improved the survival rate of scion after grafting [17].

2.1.2 Cell Wall Metabolism and Cell Expansion

The plant grows as the result of cell division, elongation and differentiation. The ascorbate, as well as several enzymes involved in ascorbate synthesis and metabolism is reported to affect the cell wall metabolism and cell growth directly or indirectly.

Ascorbate may act to regulate cellular process including cell wall metabolism by linking its biosynthesis with other cell metabolism. VTC4, the enzyme of L-galactose 1-phosphate phosphatase (GPP) for ascorbate synthesis, is shown to act as a bifunctional enzyme that affects myoinositol and ascorbate biosynthesis in plants. Myoinositol synthesis and catabolism are crucial for the production of phosphatidylinositol signaling molecules, glycerophosphoinositide membrane anchors, and cell wall pectic noncellulosic polysaccharides [18].

The ascorbate helps to eliminate free radicals involved in xylogen synthesis, regulate the polymerization of xylogen monomers and the lignification of cell wall. The exogenous monodehydroascorbate will promote the cell growth and rooting of onion [19, 20]. The ascorbate peroxidase (APX) keeps the plasticity of cell wall by reducing hydrogen peroxide [19, 20]. The ascorbate reversibly inhibits the activity of apoplast APX, prevents the conversion and secretion of free radicals into the apoplast, and regulates the lignification of plant cell wall [20, 21]. The equilibrium between ascorbate and hydrogen peroxide regulates the polymerization of xylogen monomers, and thus modulates the cell wall lignification [22].

The monodehydroascorbate is formed by oxidizing the ascorbate with AO enzyme, and then reduced by cytochrome b on the plasmalemmas, during which transmembrane electron transport promotes the cell growth [4]. Addition of monodehydroascorbate to ascorbate-loaded plasmalemma vesicles caused a rapid oxidation of the cytochrome, followed by a slower re-reduction [23, 24]. Thus, the ascorbate free radical could act as an electron acceptor to the cytochrome-mediated electron transport reaction, and eventually promote cell growth, as the transmembrane electron transport has been shown to stimulate cell growth.

The ascorbate and dehydroascorbate in the cell wall can affect the crosslinking of cell wall protein and polysaccharide, leading to loosening of the cell wall [25]. The ascorbate-induced hydroxyl radicals was shown to promote oxidative scission of plant cell wall polysaccharides [26]. It is proposed that ascorbate non-enzymically reduces O_2 to H_2O_2, and Cu^{2+} to Cu^+, and that H_2O_2 and Cu^+ react to form $^\cdot OH$, which causes oxidative scission of polysaccharide chains. Although $^\cdot OH$ radicals are often regarded as detrimental, they are so short-lived that they could act as site-specific oxidants targeted to play a useful role in loosening the cell wall, e.g. during cell expansion, fruit ripening and organ abscission.

Ascorbate acts as substrate for oxalate biosynthesis, and apoplast oxalate can influence the crosslinking of pectin and elongation of cell wall by binding Ca^{2+}. The dehydroascorbate is converted to cell wall oxalate, which binds Ca^{2+} into crystallization and regulates the Ca^{2+} level in cell wall indirectly [22]. In addition, ascorbate is a cofactor for prolyl hydroxylase that posttranslationally hydroxylates proline residues in cell wall hydroxyproline-rich glycoproteins required for cell division and expansion [25].

The cell wall-localized AO enzyme, which is involved in ascorbate oxidation and metabolism, has been implicated in control of cell growth. AO enzyme are closely related to cell expansion and cell division [27, 28]. AO enzyme activity is found mainly in plant cell wall, especially in the fast growing cells [29, 30]. High AO activity in the cell wall is correlated with areas of rapid cell expansion.

The enzyme responsible for ascorbate synthesis is also reported to be involved in cell growth. In addition to ascorbate synthesis, L-galactose dehydrogenase (GalLDH), an enzyme for ascorbate synthesis, could play an important role in the regulation of cell growth-related processes in plants. In all *SlGalLDH*-RNAi plants with reduced *SlGalLDH* transcript and enzyme activity, plant growth rate was decreased [31]. The most affected lines studied exhibited up to an 80 % reduction in SlGalLDH activity and showed a strong reduction in leaf and fruit size, mainly as a consequence of reduced cell expansion.

All these evidence indicate that ascorbate and/or enzymes involved in its synthesis and metabolism are closely associated in the process of cell wall metabolism and cell expansion.

2.1.3 Shoot Apical Meristem Formation

Ascorbate level in plants can influence the plant shoot apical meristem formation [32]. Among the transgenic tomato plants expressing *GME* gene, which is responsible for ascorbate synthesis, several co-suppression transgenic plants were identified. Interestingly, the inhibition of apical buds formation and early initiation of auxiliary buds were observed in *SlGME2* co-suppression tomato seedlings (unpublished data).

The ascorbate content in the *SlGME2* co-suppression plants with abnormal shoots was only about 1.9–15.0 % of the wild-type control. This sharp decreasing in ascorbate in the *SlGME2* co-suppression lines was found to be associated with abnormal phenotype in seedlings. The apical bud growth was retarded, leading to the absence of apical dominance and early formation of auxiliary buds (Fig. 2.1) (unpublished data).

The GME co-suppression lines with significant decreasing of ascorbate content also exhibited brown spots on the apical buds and shortened internodes (Fig. 2.1; Table 2.1). In the *GME* co-suppression tomato plants, the brownish region appeared on the apical shoots of the seedlings of 4–5 true leaf stage (Fig. 2.1). The brown sector expanded 1 week later and the apical shoots were retarded 2 weeks later. At the same time, the axillary shoots below the apical shoots were generated and grew upward instead of the apical shoots. Some of the grown-up axillary shoots were also retarded in the same pattern with that of apical buds, and replaced by the surrounding clusters of axillary buds (Fig. 2.1). This phenotype changes were consistently observed in open filed and greenhouse condition in three generations, indicating that the phenotype alteration was most likely caused by gene suppression instead of environment or tissue culture (unpublished data).

This phenotype alteration was shown to be inheritable from T_0 to T_2 generations. However, in the T_2 generation of the *GME* co-suppression lines, not all the offspring plants exhibited the shoot retardation. The abnormal growth ratio of the co-suppression line of GME2-1 and GME2-3 was 17.8 and 13.9 %, respectively. To investigate the phenotype more accurately, the two-month old seedlings were utilized to analyze the average plant height, internode length, and axillary bud number. The GME2-1 and GME2-3 lines with growth retardation has the plant height of 6.93 cm (36 % of control) and 3.80 cm (20 % of control), respectively, while the untransformed control plant was 19.17 cm high (Table 2.1; Fig. 2.2). The co-suppression lines with growth retardation of GME2-1 and GME2-3 had the internode length of 0.86 cm (45 % of the control) and 0.89 cm (46 % of the

Fig. 2.1 Abnormal apical buds in seedlings of *GME* co-suppression tomato. **a** Apical buds with brown spots in *GME* co-suppression line; **b** Increase of brown areas on buds in *GME* co-suppression line; **c** Formation of lateral buds in *GME* co-suppression line; **d** Apical buds of wild-type seedlings; **e** Lateral shoot clusters grow over the apical shoots in *GME* co-suppression lines (courtesy of Professor Zhibiao Ye and Dr. Chanjuan Zhang)

Table 2.1 The average plant height, internode length and lateral bud number of *GME* transgenic plants and the wild-type control (WT)

Line	The average plant height (cm)	The average internode length (cm)	Number of lateral buds
WT	19.17 ± 1.47	1.92 ± 0.15	0
GME2-1(S)	6.93 ± 2.19[a]	0.86 ± 0.30[a]	1–2[a]
GME2-3 (S)	3.8 ± 0.64[a]	0.89 ± 0.23[a]	2–3[a]
GME2-1 (N)	20.9 ± 1.15	1.99 ± 0.14	0
GME2-3 (N)	17.05 ± 1.94	1.69 ± 0.03	0
GME1-40 (O)	19.13 ± 1.56	1.85 ± 0.18	0
GME2-6 (O)	21.05 ± 2.03	2.05 ± 0.25	0

S denotes the offspring plants with *GME* co-suppression and abnormal phenotype. *N* represents the transgenic offspring with mild co-suppression and normal phenotype. *O* represents the overexpressing lines with normal phenotype. [a] indicates values that are significantly different from those of wild-type

Fig. 2.2 *GME* co-suppression tomato (GME2-3) exhibited shortened internode length and lower plant height (courtesy of Professor Zhibiao Ye and Dr. Chanjuan Zhang)

control), respectively. However, the co-suppression lines of GME2-1 and GME2-3 without growth retardation and the *SlGME* overexpressing lines showed similar plant height and internode length with that of wild-type plants. Among the different lines tested, the axillary buds appeared only on the co-suppression lines of GME2-1 and GME2-3 with growth retardation (Table 2.1) (unpublished data).

It should be noted that ascorbate content in the co-suppression lines of GME2-1 and GME2-3 with growth retardation was much lower than that in other mild co-suppression lines without phenotype alterations. That means the restrained ascorbate level is supposed to be associated with the plant development retardation.

On the other hand, spraying ascorbate in advance could avert the *GME* co-suppression lines from abnormal growth. The co-suppression seedlings supplemented with exogenous ascorbate did not show brown spots or abnormal growth, while the control plants (sprayed with water) eventually exhibited the abnormal phenotype. For the four leaf stage seedlings, the co-suppression line of GME2-3 were hydroponically cultured for 2 days and then sprayed with 100 mmol/L ascorbate. The mock spraying (with water) lines developed more brown regions and the apical shoots were retarded with developing axillary buds.

The ascorbate supplement, however, resulted in a limited growth of axillary buds, and restored apical shoot growth. One week after the ascorbate spraying stopped, however, the resorted the apical shoots turned brownish again with abnormal growth, probably because the exogenously supplemented ascorbate was consumed, resulting in ascorbate shortage in the seedlings.

That the co-suppression of *SlGME2* resulted in retardation of the apical shoot growth in seedlings, which could be restored by spraying ascorbate solutions, suggested the abnormal phenotype was related to extreme deficiency of ascorbate. The abnormal apical shoots appeared again shortly after the exogenous ascorbate spraying stopped possibly because supplemented ascorbate was consumed and the plants were again confronted with ascorbate shortage. Up to now, in all the ascorbate defective mutants identified in Arabidopsis, the ascorbate content in leaves was no less than 30 % of the wild-type. The extremely low content (less than 15 % of wild-type) of ascorbate in the *SlGME2* co-suppression tomato plants was possibly the threshold for plants to survive.

That is reminiscent of the previous investigation that the RNA interference of the *SlGME1* and *SlGME2* simultaneously in tomato resulted in growth retardation [33]. However, the abnormal apical shoot development was not observed in this case. The exogenous supplement of ascorbate did not restore the phenotype in seedling [33]. The mild decreasing in *SlGME1-SlGME2 RNAi* plants might not activate plant sensitivity to the ascorbate supplement.

The retarded apical shoot development in the *GME* co-suppression tomato plants may result from cell apoptosis in apical meristem as revealed by cytological analysis (unpublished data). In the *SlGME2* co-suppression plants with abnormal shoot development, the paraffin section and transmission electron microscopy observation showed that most of the parenchymatous cells in the medulla adjacent to shoot apical meristem were collapsed or apoptotic (Fig. 2.3) (unpublished data).

The seedlings of five leaves stage were utilized for the paraffin section analysis. The vertical section of the GME2-1 and GME2-3 co-suppression lines showed that the brown spots did not affect the shoot apical meristem significantly. The normal leaf primodia could be observed. However, the parenchyma cells in the co-suppression lines were unusually dyed with abnormal structure compared to those regular well-organized cells in the wild-type control (WT). At the same time, more vascular bundles were observed in the co-suppression lines, indicating the abnormal apical shoot development probably came from cell death in the medulla (Fig. 2.4). And the more vascular bundles suggested that the apical shoot meristem of co-suppression lines activated early cell differentiation (unpublished data).

The plant aerobic metabolism such as respiration and photosynthesis will produce large amount of ROS, and the surplus ROS exert damages on the protein, unsaturated fatty acid and DNA, resulting oxidation injury and function disorder in cells. Both enzymatic and non-enzymatic antioxidants are utilized in plants to scavenge the excessive ROS. The decreasing amount of ascorbate, one of the most important antioxidants, in the *SlGME2* co-suppression plants, will certainly affect the plant's ability of scavenging ROS, resulting in the accumulation of ROS and oxidization damage.

WT **GME2-3**

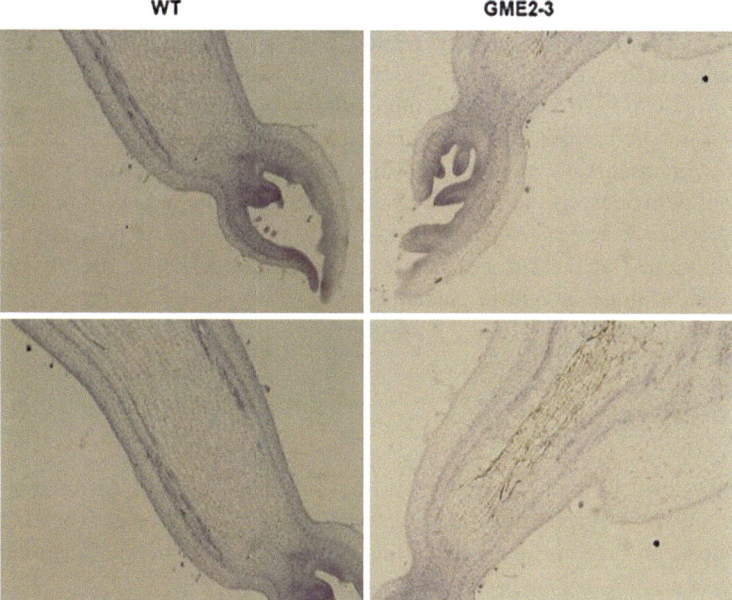

Fig. 2.3 Microsections of seedlings shoot apical meristem and stem from GME2-3 co-suppression line and control plants (WT) at five leaf stage (courtesy of Professor Zhibiao Ye and Dr. Chanjuan Zhang)

WT **GME2-3**

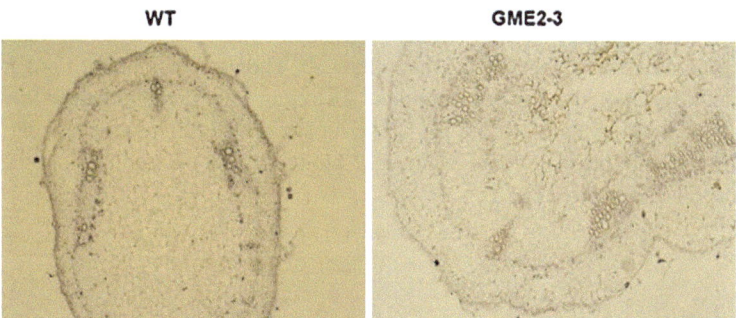

Fig. 2.4 Microsections of stem adjacent to shoot apical meristem from GME2-3 co-suppression line and control plants (WT) at five leaf stage (courtesy of Professor Zhibiao Ye and Dr. Chanjuan Zhang)

The *SlGME2* co-suppression lines showed increased sensitivity to oxidative stress. This is likely attributed to the sharply decreasing ascorbate in the co-suppression lines, since ascorbate metabolism affects the plant capacity to scavenge ROS and resistance to oxidative stress. The ROS act as cell death-inducing signal as well as xylem differentiation regulator. The increasing xylem number in the stem of tomato seedlings with *SlGME2* co-suppression was probably attributed to the accumulation of hydrogen peroxide (Fig. 2.4).

Thus the interesting co-suppression lines of *GME* presented a good demonstration that ascorbate is closely related with apical shoot development. The ascorbate promotion of shoot formation is also supported by physiological evidence. Application of ascorbate promotes shoot formation from the cut surface of tomato stems. Ascorbate levels and the activities of antioxidant enzymes in plants treated with ascorbate were higher than those in control plants. Conversely, hydrogen peroxide concentrations and malondialdehyde contents in plants treated with ascorbate were lower than those in control plants. It is supposed that cells in the calli formed at the cut surfaces re-differentiated earlier in ascorbate-treated plants than in those in control plants [34].

2.1.4 Root Development

Ascorbate level in plants is reported to be linked to development of root architecture and root response to gravity. And the ascorbate defective mutants in Arabidopsis provide efficient genetic tool to investigate ascorbate function in root development.

The root architecture is reported to be modulated in the Arabidopsis ascorbate defective mutants with moderately low (*vtc1*) or very low (*vtc2*) ascorbate content, as compared to wild-type [35]. Although the shoot development was comparable in all accessions over the first 14 d of growth, the production of primary roots was slightly different in *vtc1*, *vtc2*, and wild-type plants. The *vtc* mutants showed the antagonistic interaction between nitrate and sugar in the regulation of lateral root development that was observed in the wild-type. The *vtc2* mutant with much lower ascorbate level produced greater numbers of longer lateral roots than wild-type or *vtc1* plants at all levels of nitrate [35]. The most notable difference was that a high proportion of the primary roots of the *vtc2* plants grown on soil had lost the wild-type responses to gravity.

2.1.5 Photosynthesis

Ascorbate has proposed functions in photosynthesis as an enzyme cofactor (including synthesis of ethylene, gibberellins and anthocyanins). It has a major role in photosynthesis, acting in the Mehler peroxidase reaction with APX to regulate the redox state of photosynthetic electron carriers and as a cofactor for violaxanthin deepoxidase, an enzyme involved in xanthophyll cycle-mediated photoprotection [30].

Ascorbate accumulation in Arabidopsis leaves is increased by high light along with expression and activity of GDP-L-galactose phosphorylase (GGP, also VTC2), the enzyme responsible for ascorbate synthesis. That indicates the multiple roles of ascorbate during photosynthesis. These roles may include modulation of hydrogen peroxide and singlet oxygen, enzyme cofactor in the xanthophyll cycle and, speculatively, a photosystem II electron donor during photoinhibition [9].

Role of ascorbate in photosynthesis is also supported by the transgenic plants regulating the enzyme for ascorbate synthesis. Suppressed expression of L-galactono-1,4-lactone dehydrogenase gene (*GLDH*), the gene encoding last step enzyme for ascorbate synthesis, in rice resulted in a loss of chlorophyll, a lower Ribulose 1,5-bisphosphate carboxylase/oxygenase protein content, and a lower rate of CO_2 assimilation. As a consequence, a slower rate of plant growth and lower seed set were observed. Conversely, increasing *GLDH* expression maintained high levels of chlorophyll, Rubisco protein, and a higher rate of net photosynthesis, resulting in higher seed set [36]. These data at least indicate that the ascorbate level and/or GLDH enzyme is closely associated with plant photosynthesis and growth.

2.1.6   Regulation of Florescence

The florescence of higher plants is jointly modulated by both endogenous and exogenous regulators. The external factors include day length, illumination and temperature, while the internal factors include gibberellins etc. Recent study indicated that ascorbate participates in the physiological process of florescence regulation.

The Arabidopsis ascorbate defective mutants, *vtc1-1*, *vtc2-1*, *vtc3-1* and *vtc4-1*, exhibited earlier florescence as compared to wild-type plants regardless of long or short days [37]. However, late flowering was observed in double mutant Arabidopsis lacking both thylakoid APX and cytosolic APX [38]. Antisense suppression of *AO* gene in transgenic tobacco resulted in delayed flowering time with respect to the wild-type control under normal growth, while alteration in other phenotypes was not observed [39]. Feeding the wild-type plants of Arabidopsis with L-galactono-1,4-lactone, the direct precursor for ascorbate synthesis, resulted in five-day delay of florescence. The L-galactono-1,4-lactone feeding on transgenic Arabidopsis expressing fusion gene *LEAFY::GUS* led to postponed expression of the *LEAFY* gene in apical parts of Arabidopsis [40]. All these evidences consistently suggest that ascorbate level is potentially correlated with the flowering time. Nevertheless, whether the flowering time is associated with the ROS accumulation or ascorbate level remain undetermined.

The mechanism underlying the ascorbate mediated regulation of florescence remains elusive. Break through came in 2009 about the ascorbate regulation of Arabidopsis florescence. Expression analysis of florescence related genes in Arabidopsis ascorbate defective mutant showed that gene expression alteration coincided with the phenotype of early florescence. The expression level of genes related to florescence in ascorbate defective mutant was significantly higher than that of wild-type. The feeding of L-galactose resulted in improved ascorbate content in the both wild-type plants and the defective mutant, and delayed florescence as compared to the mock control (feeding with water). The L-galactose feeding also caused decreasing expression of genes related to flowering and photoperiodicity as compared to water feeding [37]. Double mutants combining ascorbate defective mutant *vtc* and photoperiodic or autonomous pathway mutants showed decreasing ascorbate content in Arabidopsis and delayed flowering time. Thus ascorbate might act upstream of photoperiodic and autonomous pathways to regulate the flowering time. Kotchoni et al. suggested that ascorbate act as an endogenous signal to influence the flowering time by modulating the related gene expression and metabolism process [37].

2.1.7 Regulation of Leaf Senescence

Senescence, a type of programmed cell death in plants, is initiated as the last stage during the leaf development. Both external and internal factors can trigger and aggravate leaf senescence. External factors include extreme temperature, drought, malnutrition, ozone, illumination deficiency and disease, while the internal factors affecting the leaf senescence are comprised of physiological age and developmental stages of reproductive organ.

Series of physiological and biochemical changes, such as cell structure, cell metabolism, and related gene expression, take place in plant cells during the process of senescence. The early stage of leaf senescence shows chlorophyll degradation and decreasing photosynthetic capacity due to the decreasing expression of genes related to Rubisco smaller subunit and chlorophyll a/b binding proteins, which are termed as senescence-down-regulated genes (*SDGs*). Some other genes are up-regulated in the early stage of leaf senescence, which are called senescence-associated genes (*SAGs*). The latter stage of leaf senescence shows cell peroxidation, DNA degradation and eventually disintegrated organelles.

Ascorbate is shown to influence the senescence of plants by modulating the expression of *SAGs*. Low accumulation of ascorbate accelerates senescence, while high content of ascorbate postpones the plant senescence. When detached leaves were treated in dark, leaves of ascorbate defective mutant, *vtc1*, lost chlorophyll more quickly and were induced to senesce earlier than wild-type plant leaves. Early expression of senescence-associated genes such as *SAG13*, *SAG15* were observed in ascorbate mutant of *vtc1* [41]. When supplemented with exogenous ascorbate, however, the expression levels of *SAG13* of the mutant were restored to the levels of wild-type. Additionally, the expression abundance of some atypical *SAGs* such as *PR-1*, *PR-2* and *PR-5*, in the mutants of *vtc1* and *vtc2* was higher than that of wild-type control.

The ROS are shown to promote the expression of senescence-associated genes, during which ascorbate is supposed to be involved in the transcriptional regulation of *SAGs*. In the leaves treated with silver nitrate, a ROS generating reagent, the *LSC54* and *LSC94* were significantly up-regulated until the subsequent treatment with ascorbate [42]. Expression of the *LSC54* and *LSC94* gene has been shown previously to increase during leaf senescence and cell death. Supplement of ascorbate will reduce the ROS accumulation, alleviate oxidative damage to photosynthetic tissue, and consequently delay the process of senescence. During this process, the ascorbate may regulate leaf senescence by modulating the ROS level and/or the expression of *SAGs*.

The role of ascorbate in senescence regulation is also supported with physiological evidence. At later stages of plant development, the *vtc* rosettes were smaller than those of the wild-type and the leaves showed intracellular structural changes that are consistent with programmed cell death (PCD). PCD symptoms such as nuclear chromatin condensation, the presence of multivesicular bodies, and extensive degradation and disorganization of the grana stacks were observed in 8-week-old *vtc2* leaves and in 10-week-old *vtc1* leaves [35].

The *GMP* gene was reported to impose influences on the plant senescence. Decreasing ascorbate content resulted in early senescence in transgenic potato with antisense expression of GDP-D-mannose pyrophosphorylase gene (*GMP*), a gene in ascorbate synthesis pathway [43]. Brown spots appeared on the stem and leaves of *GMP* antisense transgenic plants and spread from bottom to top 10 weeks after transferring to soil. Those antisense transgenic lines with the most significant decrease of GMP enzyme activity and the most serious symptoms withered and died three months after transferring to soil, while the wild-type plants did not start to senesce [43]. On the contrary, *GMP* overexpression in potato resulted in improved ascorbate content and increasing ratio of ascorbate/dehydroascorbate in both leaves and tubers of transgenic potato plants. Both pigment content and photosynthetic rate were much higher in transgenic plants than that in wild-type plants. *GMP* overexpressing plants showed a distinguishable change in phenotype and delayed senescence [44].

Consistently, in the *GMP* RNAi-suppressed tomato plants, similar early senescence phenotype was observed. The senescence in the *GMP* suppressed tomato plants happened even earlier than that of transgenic potato plants [43]. In the *GMP* RNAi transgenic tomato seedlings, brown spots were formed on young leaves and stem, resulting in rapid etiolation and abscission of the bottom leaves. As the plants developed, the early senescence and yellowing of leaves expanded from bottom to the upper parts of the plants, and eventually most of the leaves withered (unpublished data).

Although the decreasing ascorbate accumulation is connected with early senescence in plants, the *SlGME2* suppressed transgenic plants did not generate brownish spots or early senescence on leaves. This suggests that the early senescence in the *SlGMP* RNAi transgenic plants is probably caused by the ascorbate deficiency as well as the insufficient product of GMP enzyme catalyzing, GDP-D-mannose, which is vital for various cellular process. GDP-D-mannose is required for the biosynthesis of the glucomannan and galactomannan of the hemicellulose polymers in plant cell wall, the GDP-L-trehalose and glycoprotein in the cell wall, and the O-linked glycoprotein and N-linked glycoprotein in the protein glycosylation. Microarray analysis showed that several genes related to cell wall synthesis and protein glycosylation were up-regulated in the break and ripening stage fruits of transgenic tomato overexpressing *SlGMP*, indicating that the biological processes other than ascorbate biosynthesis were altered by *GMP* overexpression [32]. Thus the early senescence and brownish spots on the *GMP* suppressing transgenic tomato plants might be the consequence of down-regulated gene expression related to cell wall synthesis and/or protein glycosylation. The mechanism underlying the interaction between ascorbate synthesis and leaf senescence invites further investigation.

Ascorbate is the cofactor for enzymes involved in biosynthesis of GA, ABA and ethylene. Emerging evidences indicate that ascorbate together with various phytohormones regulates the process of senescence. The phytohormones of ABA and ethylene promote senescence while GA prevents senescence. Thus, ascorbate possibly regulate the plant senescence either by modulating the ROS accumulation or influencing the signal pathway of the phytohormones, such as GA, ABA and ethylene [40]. Thus, in addition to acting simply as an antioxidant in the apoplastic space, ascorbate appears to be involved in a complex phytohormone-mediated signalling network that links together ROS responses and the onset of senescence [45].

Taken together, the redox status as well as the concentration of ascorbate is involved in the regulation of plant senescence [44]. The decreasing ascorbate content usually results in early senescence, which is also consistent with the fact that the ascorbate is negatively involved in the regulation of flowering time.

2.2 The Cofactors for Enzyme Activity

The ascorbate acts as cofactors to regulate the series of enzyme activity and facilitate enzymatic reactions [46]. Ascorbate interacts with enzymes having either monooxygenase or dioxygenase activity. Ascorbate is an essential cofactor in reactions catalyzed by Cu^{+}-dependent monooxygenases and Fe^{2+}-dependent dioxygenases [22, 25]. In the plants, ascorbate is utilized as coenzyme for the Fe dioxygenase to participate in the post-translational modification of the cell wall protein. The dehydroascorbate can interact with the side chain of lysine and arginine in the enzyme and prevent protein crosslinking [47]. Ascorbate is also reported to modulate the ferritin-mediated iron uptake and release in plants [48]. The ascorbate keeps the reduced state of the metal ions in the enzymatic reaction center, and promotes the enzymatic activity [22].

Ascorbate is also a cofactor for some hydroxylase enzymes (e.g. prolyl hydroxylase) and violaxanthin deepoxidase. Ascorbate acts as prothetic group for prolyl hydroxylase and lysyl hydroxylase, catalyzing the synthesis of hydroxylysine and hydroxyproline [25]. And the violaxanthin deepoxidase enzyme links ascorbate to the photoprotective xanthophyll cycle [4]. As the cofactor of violaxanthin deepoxidase, ascorbate is thus involved in the pigment biosynthesis for photoprotection [22].

Ascorbate is shown to be enzyme cofactors involving synthesis of ethylene, GA and anthocyanins. Ascorbate is the coenzyme for 1-aminocyclopropane-1-carboxylate oxidase (ACC oxidase) and GA2-oxidase and thus involved in the biosynthesis of phytohormones of ethylene and GA [4]. In addition, the ascorbate can strongly activate myrosinase, a family of enzymes involved in plant defense against herbivores.

2.3 Plant Antioxidation Capacity

In both plant and animal metabolism, the biological functions of ascorbate are centered on the antioxidant properties of this molecule. Considerable evidence has been accruing in the last three decades of the importance of ascorbate in protecting not only the plant from oxidative stress, but also mammals from various chronic diseases that have their origins in oxidative stress. In plants, ascorbate is the most important antioxidant and, in association with other components of the antioxidant system, protects plants against oxidative damage resulting from aerobic metabolism, photosynthesis and a range of pollutants.

The hypersensitivity of some of the *vtc* mutants to ozone and UV-B radiation, the rapid response of APX expression and ascorbate level to oxidative stress, and the properties of transgenic plants with altered APX activity all support an important role for ascorbate in protecting plants against oxidative stress [4, 30]. In the *vtc-1* mutant of Arabidopsis, which had an ascorbate deficiency in chloroplasts of ca. 60 %, although low ascorbate did not cause oxidative stress in optimal growth conditions, it increased malondialdehyde levels in chloroplasts by ca. 60 %, and reduced α-tocopherol (vitamin E) and β-carotene by ca. 85 and 40 %, respectively, in water-stressed mutants, showing that ascorbate contributes to the protection of thylakoid membrane lipids from oxidation in stressed plants [49].

The ascorbate level as well as activities of the biosynthesis related enzymes in plants responds rapidly to oxidative stress. The content of ascorbate increased in plants grown in polluted air together with the enzymatic activities of dehydro-ascorbate reductase (DHAR) involved in its recycling. However, the APX activity remained unchanged in this case. The data obtained support the hypothesis that the glutathione-ascorbate cycle is stimulated in removing the hydrogen peroxide produced under oxidative stress [50]. APX, the enzyme responsible for ascorbate oxidation, is an important enzymatic ROS scavenger. APX enzyme activities are shown to be correlated with oxidation tolerance [51]. APX activity increased after ozone treatment and this effect was stronger if the plants were pre-treated with anti-ozonant ethylenediurea prior to O_3-exposure. The ozone exposure of plants stimulates the APX, whose enzymatic activity enhancement at the apoplastic and cytosolic levels may act as a biochemical defence against ozone damage [52].

Scavenging the ROS in plants is one of the major tasks of ascorbate. The ROS is generated during photosynthesis and respiration, as well as in various stress such as drought, extreme temperature, salt, ultraviolet etc. Excessive ROS will do harm to lipid, nuclear acid, and protein, resulting in oxidative damages, physiological disorder, and eventually premature senescence and cell death [53]. Ascorbate is an effective radical scavenger to eliminate the ROS in plants. Ascorbate reacts with ROS generating monodehydroascorbate and dehydroascorbate. The dehydro-ascorbate is decomposed into tartaric acid and ketosuccinic acid. The monode-hydroascorbate and dehydroascorbate are reduced into ascorbate by MDHAR enzyme using NADPH or glutathione as reducer. ROS can be consumed through the resultant ascorbate–glutathione cycle. For activity, ascorbate must be in the fully reduced state. Therefore, both the rate of ascorbate synthesis and recycling via dehydroascorbate and monodehydroascorbate reductases are critical in the process of ROS scavenging [45].

Ascorbate is a well-known antioxidant and cellular reductant with an intimate and complex role in the response of plants to ozone. It is clear from a number of studies that sensitivity to ozone is correlated with total ascorbate levels, and that a first line of defence against the ROS generated in the apoplastic space by ozone is ascorbate. The malondialdehyde is the product of membranous peroxidation. The malondialdehyde content represents the damage extent of the plant cells or the membrane permeability. The ascorbate can inhibit the membranous peroxidation, protect the cell from damaging, and delay the cell senescence by decreasing the production of malondialdehyde and stabilizing the cytomembrane structure.

The ascorbate plays a pivotal role in scavenging the ROS generated in the process of photosynthesis. The leaf ascorbate content is usually higher than that in other tissues of plants. This, from an evolutional point of view, will help plant to defense against the ROS generating from photosynthesis. The leaf apoplast con-tains millimolar amounts of ascorbate that protects the plasmalemma against oxidative damage due to the process of photosynthesis [5]. High concentration of ascorbate is accumulated in the chloroplast, while chloroplast is in lack of catalase (CAT). Thus the hydrogen peroxide produced in the photosystem I has to be scavenged by APX [54]. The ascorbate is demonstrated to act as an electron donor to photosystem I in light-induced electron transport. It was found that ascorbate, at physiological concentrations, rapidly reduced photooxidized reaction center chlorophyll of photosystem I [55]. The oxidized ascorbate of monodehydro-ascorbate is also the direct electron acceptor in the photosystem I [56]. The levels and redox state of ascorbate modify the pattern of modulation of photosynthesis by mitochondrial metabolism [57]. Ascorbate is also co-factor of violaxanthin deepoxidase, which is a key enzyme of the xanthophyll cycle converting viola-xanthin to cryptoxanthin on the thylakoid membrane. Thus the ascorbate plays an important role in consuming the excessive light energy and protecting photosyn-thesis for plants.

At a physiological level, the best-studied phenomena involving ascorbate is its participation in an oxygen scavenging pathway in the chloroplast known as the ascorbate–glutathione cycle [6]. However, either cytosolic ascorbate–glutathione cycle or chloroplastic ascorbate–glutathione cycle is primarily utilized for the detoxification of hydrogen peroxide may vary with plant species [58].

The antioxidative role of ascorbate also resides in its capacity to maintain quality of plant products, e.g. fruit and vegetable. The ascorbate shows inhibitive effect on the polyphenol oxidase (PPO), which oxidizes diphenols to quinones resulting in browning reactions in many wounded horticultural fruits and vegetables. The ascorbate can inhibit the activity of PPO enzyme by reducing the quinones and its derivatives into phenolic compounds or prevent the spontaneous polymerization of quinones into pigment substances. The ascorbate helps to keep another important antioxidant, vitamin E, in the reduced state.

Ascorbate-dependent detoxification of hydrogen peroxide is also shown to be associated with guaiacol-type peroxidases as well as ascorbate peroxidase, suggesting that ascorbate is the natural substrate for many types of peroxidase in situ and not just the ascorbate-specific peroxidases [59, 60]. The ascorbate-dependent destruction of hydrogen peroxide in the more acidic cellular compartments such as the vacuole may be an important function of such non-specific peroxidases [61].

2.4 Heavy Metal Evacuation and Detoxification

Ascorbate is supposed to act in the plant defence against heavy metal stress. Ascorbate may help plants to eliminate the heavy metals or ROS generated by this stress.

Ascorbate is shown to promote the $Hg°$ emission originated from Hg^{2+} uptake by the roots. Homogenates of barley leaves added to dissolved Hg^{2+} induced a powerful volatilization at alkaline but not at acidic pH. The same pH dependence and emission kinetic together with the highest reduction capacity was observed for ascorbate as compared to other phytoreductants. The electrochemical potentials of the reactions involved suggest an electron transfer from NADPH via glutathione and ascorbate to Hg^{2+}. The results showed plants transfer reduction equivalents via ascorbate to reduce Hg^{2+} ions, thus counteracting mercury toxicity by volatilizing the metal. This effect appears to be assisted by other light-dependent processes such as transpiration and ascorbate synthesis [62].

In an initial exposure to heavy metals (Cd^{2+}, Pb^{2+} and Hg^{2+}), the ascorbate levels increases in plants, suggesting that the ascorbate respond actively to evacuate heavy metals or detoxify the ROS derived from heavy metal stress [63].

2.5 Role in Stress Defense

Environmental stresses include both biotic and abiotic stresses. Abiotic stress is one of the limiting factors for crop production. Biotic stress comes from insects and diseases, while abiotic stress is comprised of drought, salt, cold, high temperature, ultraviolet, and heavy metals etc. These environmental stress factors cause redox imbalance and oxidative damage in plants. The ROS, such as superoxide anion (O^{2-}), hydroxyl free radical ($^{\cdot}OH$), hydrogen peroxide (H_2O_2) and singlet oxygen, are generated under various abiotic and biotic stresses. ROS can cause the peroxidation of plasmalemma, DNA mutation, protein denaturation, and eventually cell death [64]. Thus it is quite indispensable to scavenge the surplus ROS for plant growth and development.

In recent years, the role of the plant antioxidant system for the stress response has become a research focus of plant stress physiology. Generally it is regarded that plant injury caused by stress comes from antioxidant system decreasing and membrane lipid peroxidation. A variety of antioxidant enzymes such as superoxide dismutase (SOD), peroxidase (POD), APX and glutathione reductase (GR), etc., as well as non-enzymatic antioxidants such as ascorbate, glutathione, vitamin E and carotenoids play important roles in plant responses to stresses [29, 65]. These enzymatic antioxidants and non-enzymatic antioxidant are utilized to eliminate excess ROS in plant cells.

Ascorbate has been shown as an efficient ROS scavenger in the process of plant defense to various stresses [3, 28]. Ascorbate participates directly in eliminating ROS as electron donor, and the enzymes involved in ascorbate metabolism also act positively in plant defense against various stress. The ascorbate has great affinity with anionic peroxidase, which detoxifies the hydrogen peroxide during the processes of peroxidation of ascorbate to defend against peroxidative stress [66]. As the most important antioxidant in plant cells, one of the key functions of ascorbate is to protect the chloroplast from oxidative damage.

Ascorbate can scavenge the ROS directly or indirectly, which is generated during various physiological process like photosynthesis, oxidization and metabolism, and stress response, and thus help plants to overcome the oxidative damage and survive [30]. Ascorbate in apoplast is considered to be involved in signal perceiving and transduction from external environment [67, 68]. The reduced state of ascorbate as well as the oxidized state of dehydroascorbate act as signals regulating the interaction between plant and stresses and confer resistance to stresses, such as ozone, drought and pathogen [3, 69, 70]. Apoplastic ascorbate levels and redox state has an important role in plant responses to environmental stress. Apoplastic ascorbate may be involved in the protection of plant plasmalemma against environmental stress caused by oxidative damage, especially when the plants subjected to ozone or other atmospheric pollution stress.

In living cells, ascorbate redox system is composed of reduced ascorbate, semi-reduced (monodehydroascorbate) and oxidized ascorbate (dehydroascorbate). In the intracellular environment, a number of enzymatic reactions allow the ascorbate pool to maintain at a considerable reduced state, while in the extracellular environment, redox state of ascorbate is more dependent on plant species and the physiological status. Studies have shown that the ascorbate level, the redox state (ascorbate/dehydroascorbate ratio) of ascorbate and enzyme activity involved in ascorbate biosynthesis and metabolism are related to series of environmental stress responses.

Improvement of ascorbate content in plants will increase plant stress tolerance, while decreasing ascorbate content will result in stress sensitivity of plants. Ascorbate-deficient (*vtc*) mutants tend to be smaller, more sensitive to abiotic stresses and more resistant to biotrophic pathogens. The ascorbate defective mutant in Arabidopsis, *vtc1*, showed more sensitivity to ozone, sulfur dioxide, and ultraviolet. These evidence shows that plant stress tolerance is associated with the ascorbate level in plants [71]. Grain soaking in ascorbate could also counteract the adverse effects of salinity on the seedlings [72]. That plants with higher leaf ascorbate concentrations had higher leaf NO_2 uptake rates, suggest that leaf capacity for the scavenging of NO_X ($NO + NO_2$) by ascorbate may explain the variation in the ability of plants to absorb atmospheric nitrogen oxides [73]. The ascorbate content in plants increased in response to excess level of Zn and high irradiance stress, indicating that ascorbate is one of the effective defense mechanism against stresses in plants [74]. Salt stress can lead to significantly reduced ascorbate content in wheat. However, the decreasing extent of ascorbate in salt tolerant varieties is less than that in the salt-sensitive varieties. Protein profiles under stress indicate a positive role of ascorbate in the alleviation of the damage effects induced by abiotic stress [75].

Evidence shows it is feasible to improve plant stress tolerance by regulating ascorbate synthesis and recycling to increase ascorbate accumulation. Overexpressing galacturonate reductase gene from strawberry and L-gulono-1, 4-lactone oxidase gene from rat in potato resulted in improved ascorbate synthesis and higher survival rate under oxidization and salt treatment [76, 77]. Overexpressing the ascorbate recycling gene, *DHAR*, led to enhanced resistance to various stresses in transgenic plants. Overexpressing the *DHAR* gene can increase plant defense capacity to ozone in transgenic plants [78]. Ectopic expression of human *DHAR* gene in tobacco resulted in improved resistance to cold and salt stress [79]. Overexpressing rice *DHAR* gene in Arabidopsis improved the plant resistance to salt stress [80]. Overexpressing the cytosolic *DHAR* gene from Arabidopsis in tobacco resulted in enhanced resistance to drought, ozone and aluminium [81, 82]. That overexpression of *DHAR*, but not of *MDHAR*, confers aluminium tolerance, indicates the maintenance of a high level of reduced state ascorbate is important to aluminium tolerance in plants [82].

The redox state of the ascorbate influences the abiotic resistance in plants. The ROS is formed in apoplastic regions, where ascorbate is considered to play an important role in plant defense against oxidation and injury. The AO enzyme is located in apoplast. Overexpressing the melon *AO* gene in tobacco resulted in oxidation of the ascorbate in apoplast. The reduced state of ascorbate accumulated at low level in apoplast, with altered redox status. The increasing dehydroascorbate/ascorbate ratio resulted in extreme sensitivity to ozone in *AO* transgenic plants [68]. That altered stomatal dynamics was observed in *AO* overexpressing tobacco plants suggests AO enzyme or dehydroascorbate may be closely linked to stress response, as control of stomatal aperture is of paramount importance for plant adaptation to the surrounding environment [69]. Because the ascorbate is also directly involved in cell wall elongation and lignification, maintaining high level and high redox state of apoplastic ascorbate is physiologically important for the plants during the environmental stress.

Under ozone stress conditions, the total ascorbate content of kidney bean leaves of ozone-tolerant varieties is higher than that of the sensitive varieties, and the tolerant species maintains relatively high levels of ascorbate pool. In addition, the content of apoplastic ascorbate of tolerance varieties is much higher than that of the sensitive cultivar. In the ozone tolerant varieties, the ascorbate/(ascorbate + dehydroascorbate) ratio is relatively high, suggesting that plant ozone tolerance is closely related to apoplastic ascorbate level and its redox state. Since the apoplast does not have most of the enzymes involved in the ascorbate–glutathione cycle, apoplastic ascorbate levels and its redox state further demonstrates unique ascorbate transport mechanism in plants.

The ascorbate–glutathione cycle constitutes one of the most important antioxidant systems in plants. In the ascorbate–glutathione cycle, the ascorbate and the glutathione are utilized as reducers and recycled through consuming the ATP and NAD(P)H [29], in which four enzymes, APX, MDHAR, DHAR and GR, are involved. The ascorbate–glutathione cycle enzymes are activated for the dissipation of excess excitation energy in the photosynthesis. Haem-containing enzymes include peroxidase and catalase(CAT) distributed among prokaryotes and eukaryotes and play a vital role in hydrogen peroxide detoxification [83]. The modulation of the ascorbate–glutathione cycle together with coordinated antioxidant activity involving increased activities of superoxide dismutase (SOD) and CAT, allowed plants to cope with oxidative stress [84].

As a player in the defence against ROS, the role of APX in plant stress response has been extensively studied at the biochemical and molecular level. APX, an enzyme scavenging hydrogen peroxide, is located in various organelles, such as cytoplasm, chloroplast, mitochondria and peroxisome [85]. The activity of APX, the ascorbate metabolism related enzyme, is triggered by abiotic stress. The cytoplasm-located APX enzyme, cAPX, is considered to play a crucial role in stress response of plants. The *cAPX* expression level increased significantly in leaves of pea and spinach after stress treatment of high light, drought, high temperature, and oxidation [86, 87]. The two *APX* genes in rice, *OsAPX1/2*, were up-regulated in response to injury, salicylic acid, abscisic acid and hydrogen peroxide [88]. Overexpressing cytosolic *APX* in tomato increased the plant resistance to abiotic stresses, such as ultraviolet, high temperature, cold and salt [89]. And the enzyme activity of APX was stimulated by ethylene, leading to increased resistance against ozone and other oxidative stress [90]. Interestingly, although the transcripts levels of *APX* were almost identical in control or salt-grown radish plants in both leaves and roots, the activity of APX enzyme was enhanced by the salt treatment in both leaves and roots [91], suggesting that the salt-induced APX activity is probably the consequence of post-transcriptional events. The drought acclimated leaves of wheat exhibited systematic increase in the activities of hydrogen peroxide scavenging enzymes particularly APX and CAT, and maintenance of ascorbate redox pool by efficient function of APX enzyme [92].

Fig. 2.5 Leaf discs of wild-type (WT) and transgenic tomato with *GME2* co-suppression (GME2-1 and GME2-3) treated with 100 μmol/L paraquat (courtesy of Professor Zhibiao Ye and Dr. Chanjuan Zhang)

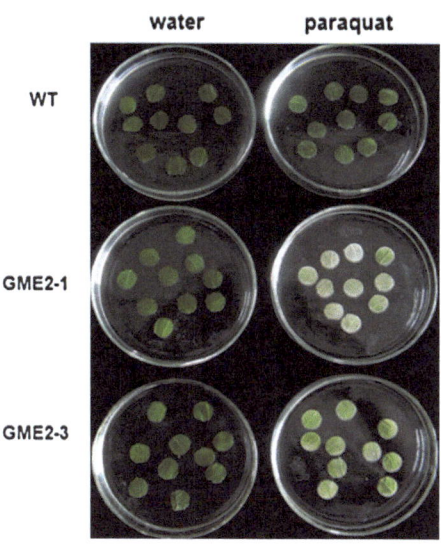

The enzymes for ascorbate synthesis also affect the plant stress responses, possibly by regulating the ascorbate accumulation. *GME* co-suppression affects the tomato response to oxidative stress. Leaf discs of two month seedlings were treated with distilled water or 100 μmol/L paraquat solution for 18 h and the chlorophyll content was investigated. Oxidation treatment of the *GME* co-suppression lines resulted in serious etiolation of leaf discs. The chlorophyll content of the transgenic plants in the water treatment was significantly lower than that of the wild-type control. The paraquat treatment of the leaf discs resulted in much more decreasing in the chlorophyll content in co-suppression lines than that in the wild-type control (Fig. 2.5). Experimental results showed that *SlGME2* co-suppression resulted in significantly decreasing ascorbate and chlorophyll content, as well as higher sensitivity to oxidative stress (unpublished data).

On the other hand, overexpression of *SlGME1* and *SlGME2* resulted in improved resistance to oxidation, chilling injury, and high salt in tomato [32]. Gene chip hybridization was utilized to analyze the gene expression profiles of *SlGME1* overexpressing tomato at the break stage fruit. The results showed that the genes related to stress resistance, lipid metabolism, flavonoid metabolism were significantly up-regulated. Several differentially expressed genes were both observed in the *SlGME1* and *SlGME2* overexpressing tomato with the similar expression profile [32].

The ascorbate is also involved in responses to biotic stress in plants. Activation of a defence gene-signalling network by both ozone and pathogens is influenced by the level of ascorbate. The ascorbate metabolism is shown to response dynamically to pathogenic bacteria infection [93]. And the ascorbate defective mutant, *vtc1*, activates numerous defense genes including those that encode pathogenesis-related proteins [3]. Interestingly, salicylic acid (SA)-deficient plants adapt to RNA virus infections better with a lighter symptom and less ROS accumulation. This symptom alleviation is supposed to attributed to the higher ascorbate content in SA-deficient plants [94]. Early and high-dose ascorbate treatment alleviates the symptom, and eventually inhibits virus replication. ROS eliminators could not imitate the effect of ascorbate, and could neither alleviate symptom nor inhibit virus replication, which indicates ascorbate may help plant defense via a unique machinery rather than the hydrogen peroxide signal [94].

Abundance of ascorbate in plants also influences their susceptibility to insect feeding. These effects may be mediated by ascorbate roles as an essential dietary nutrient, as an antioxidant in the insect midgut, or as a substrate for plant-derived ascorbate oxidase, which can lead to generation of toxic ROS. In addition, ascorbate can influence the efficacy of plant defenses such as myrosinases and tannins, and alter insects' susceptibility to natural enemies [95].

References

1. Iqbal Y, Ihsanullah I, Shaheen N, Hussain I (2009) Significance of vitamin C in plants. J Chem Soc Pak 31:169–170
2. Nishikimi M, Fukuyama R, Minoshima S, Shimizu N, Yagi K (1994) Cloning and chromosomal mapping of the human nonfunctional gene for l-gulono-gamma-lactone oxidase, the enzyme for L-ascorbic acid biosynthesis missing in man. J Biol Chem 269:13685–13688
3. Pastori GM, Kiddle G, Antoniw J, Bernard S, Veljovic-Jovanovic S, Verrier PJ, Noctor G, Foyer CH (2003) Leaf vitamin C contents modulate plant defense transcripts and regulate genes that control development through hormone signaling. Plant Cell 15:939–951
4. Smirnoff N (1996) The function and metabolism of ascorbic acid in plants. Ann Bot 78: 661–669
5. Horemans N, Foyer CH, Asard H (2000a) Transport and action of ascorbate at the plant plasma membrane. Trends Plant Sci 5:263–267
6. Horemans N, Foyer CH, Potters G, Asard H (2000b) Ascorbate function and associated transport systems in plants. Plant Physiol Biochem 38:531–540
7. Loewus FA, Loewus MW (1987) Biosynthesis and metabolism of ascorbic-acid in plants. Crit Rev Plant Sci 5:101–119
8. Loewus FA (1999) Biosynthesis and metabolism of ascorbic acid in plants and of analogs of ascorbic acid in fungi. Phytochemistry 52:193–210
9. Smirnoff N (2011) Vitamin C: the metabolism and functions of ascorbic acid in plants. Adv Bot Res 59:107–177
10. Stevens R, Buret M, Duffe P, Garchery C, Baldet P, Rothan C, Causse M (2007) Candidate genes and quantitative trait loci affecting fruit ascorbic acid content in three tomato populations. Plant Physiol 143:1943–1953
11. Cordoba F, Gonzalezreyes JA (1994) Ascorbate and plant-cell growth. J Bioenerg Biomembr 26:399–405
12. Kerk NM, Feldman LJ (1995) A biochemical-model for the initiation and maintenance of the quiescent center—implications for organization of root-meristems. Development 121: 2825–2833
13. Potters G, Horemans N, Caubergs RJ, Asard H (2000) Ascorbate and dehydroascorbate influence cell cycle progression in a tobacco cell suspension. Plant Physiol 124:17–20
14. Arrigoni O (1994) Ascorbate system in plant development. J Bioenerg Biomembr 26: 407–419

15. Veljovic-Jovanovic SD, Pignocchi C, Noctor G, Foyer CH (2001) Low ascorbic acid in the *vtc-1* mutant of Arabidopsis is associated with decreased growth and intracellular redistribution of the antioxidant system. Plant Physiol 127:426–435
16. Tabata K, Oba K, Suzuki K, Esaka M (2001) Generation and properties of ascorbic acid-deficient transgenic tobacco cells expressing antisense RNA for L-galactono-1,4-lactone dehydrogenase. Plant J 27:139–148
17. Johkan M, Oda M, Mori G (2008) Ascorbic acid promotes graft-take in sweet pepper plants (*Capsicum annuum* L.). Scientia Hortic Amsterdam 116:343–347
18. Torabinejad J, Donahue JL, Gunesekera BN, Allen-Daniels MJ, Gillaspy GE (2009) VTC4 is a bifunctional enzyme that affects myoinositol and ascorbate biosynthesis in plants. Plant Physiol 150:951–961
19. Hidalgo A, Gonzalezreyes JA, Navas P (1989) Ascorbate free-radical enhances vacuolization in onion root-meristems. Plant Cell Environ 12:455–460
20. deCabo RC, GonzalezReyes JA, Cordoba F, Navas P (1996) Rooting hastened in onions by ascorbate and ascorbate free radical. J Plant Growth Regul 15:53–56
21. Takahama U (1993). Regulation of peroxidase-dependent oxidation of phenolics by ascorbic-acid—different effects of ascorbic-acid on the oxidation of coniferyl alcohol by the apoplastic soluble and cell wall-bound peroxidases from epicotyls of *Vigna angularis*. Plant Cell Physiol 34:809–817
22. Davey MW, Van Montagu M, Inze D, Sanmartin M, Kanellis A, Smirnoff N, Benzie IJJ, Strain JJ, Favell D, Fletcher J (2000) Plant L-ascorbic acid: chemistry, function, metabolism, bioavailability and effects of processing. J Sci Food Agric 80:825–860
23. Horemans N, Asard H, Caubergs RJ (1994) The role of ascorbate free-radical as an electron-acceptor to cytochrome b-mediated trans-plasma membrane electron-transport in higher-plants. Plant Physiol 104:1455–1458
24. Asard H, Horemans N, Caubergs RJ (1995) Involvement of ascorbic-acid and a b-type cytochrome in plant plasma-membrane redox reactions. Protoplasma 184:36–41
25. Padh H (1990) Cellular functions of ascorbic-acid biochemistry and cell biology. Biochimie et Biologie Cellulaire 68:1166–1173
26. Fry SC (1998) Oxidative scission of plant cell wall polysaccharides by ascorbate-induced hydroxyl radicals. Biochemistry Journal 332:507–515
27. Kato N, Esaka M (1999) Changes in ascorbate oxidase gene expression and ascorbate levels in cell division and cell elongation in tobacco cells. Physiol Plant 105:321–329
28. Pignocchi C, Fletcher JM, Wilkinson JE, Barnes JD, Foyer CH (2003) The function of ascorbate oxidase in tobacco. Plant Physiol 132:1631–1641
29. Noctor G, Foyer CH (1998) Ascorbate and glutathione: keeping active oxygen under control. Ann Rev Plant Physiol Plant Mol Biol 49:249–279
30. Smirnoff N, Wheeler GL (2000) Ascorbic acid in plants: biosynthesis and function. Crit Rev Plant Sci 19:267–290
31. Alhagdow M, Mounet F, Gilbert L, Nunes-Nesi A, Garcia V, Just D, Petit J, Beauvoit B, Fernie AR, Rothan C, Baldet P (2007) Silencing of the mitochondrial ascorbate synthesizing enzyme L-galactono-1,4-lactone dehydrogenase affects plant and fruit development in tomato. Plant Physiol 145:1408–1422
32. Zhang CJ, Liu JX, Zhang YY, Cai XF, Gong PJ, Zhang JH, Wang TT, Li HX, Ye ZB (2011) Overexpression of *SlGMEs* leads to ascorbate accumulation with enhanced oxidative stress, cold, and salt tolerance in tomato. Plant Cell Rep 30:389–398
33. Gilbert L, Alhagdow M, Nunes-Nesi A, Quemener B, Guillon F, Bouchet B, Faurobert M, Gouble B, Page D, Garcia V, Petit J, Stevens R, Causse M, Fernie AR, Lahaye M, Rothan C, Baldet P (2009) GDP-D-mannose 3,5-epimerase (GME) plays a key role at the intersection of ascorbate and non-cellulosic cell-wall biosynthesis in tomato. Plant J 60:499–508
34. Johkan M, Mori G, Mitsukuri K, Mishiba K, Morikawa T, Imahori Y, Oda M (2008) Effect of ascorbic acid on in vivo organogenesis in tomato plants. J Hortic Sci Biotechnol 83:624–628

35. Olmos E, Kiddle G, Pellny TK, Kumar S, Foyer CH (2006) Modulation of plant morphology, root architecture, and cell structure by low vitamin C in *Arabidopsis thaliana*. J Exp Bot 57: 1645–1655
36. Liu YH, Yu L, Wang RZ (2011) Level of ascorbic acid in transgenic rice for l-galactono-1, 4-lactone dehydrogenase overexpressing or suppressed is associated with plant growth and seed set. Acta Physiol Plant 33:1353–1363
37. Kotchoni SO, Larrimore KE, Mukherjee M, Kempinski CF, Barth C (2009) Alterations in the endogenous ascorbic acid content affect flowering time in Arabidopsis. Plant Physiol 149:803–815
38. Miller G, Suzuki N, Rizhsky L, Hegie A, Koussevitzky S, Mittler R (2007) Double mutants deficient in cytosolic and thylakoid ascorbate peroxidase reveal a complex mode of interaction between reactive oxygen species, plant development, and response to abiotic stresses. Plant Physiol 144:1777–1785
39. Yamamoto A, Bhuiyan NH, Waditee R, Tanaka Y, Esaka M, Oba K, Jagendorf AT, Takabe T (2005) Suppressed expression of the apoplastic ascorbate oxidase gene increases salt tolerance in tobacco and Arabidopsis plants. J Exp Bot 56:1785–1796
40. Barth C, De Tullio M, Conklin PL (2006) The role of ascorbic acid in the control of flowering time and the onset of senescence. J Exp Bot 57:1657–1665
41. Barth C, Moeder W, Klessig DF, Conklin PL (2004) The timing of senescence and response to pathogens is altered in the ascorbate-deficient Arabidopsis mutant vitamin c-1. Plant Physiol 134:1784–1792
42. Navabpour S, Morris K, Allen R, Harrison E, A-H-Mackerness S, Buchanan-Wollaston V (2003) Expression of senescence-enhanced genes in response to oxidative stress. J Exp Bot 54:2285–2292
43. Keller R, Springer F, Renz A, Kossmann J (1999) Antisense inhibition of the GDP-mannose pyrophosphorylase reduces the ascorbate content in transgenic plants leading to developmental changes during senescence. Plant J 19:131–141
44. Lin LL, Shi QH, Wang HS, Qin AG, Yu XC (2011) Over-expression of tomato GDP-Mannose pyrophosphorylase (GMPase) in potato increases ascorbate content and delays plant senescence. Agric Sci Chin 10:534–543
45. Conklin PL, Barth C (2004) Ascorbic acid, a familiar small molecule intertwined in the response of plants to ozone, pathogens, and the onset of senescence. Plant Cell Environ 27:959–970
46. Arrigoni O, DeGara L, Paciolla C, Evidente A, dePinto MC, Liso R (1997) Lycorine: A powerful inhibitor of L-galactono-gamma-lactone dehydrogenase activity. J Plant Physiol 150:362–364
47. Fry SC (1986) Cross-linking of matrix polymers in the growing cell walls of angiosperms. Ann Rev Plant Physiol Plant Mol Biol 37:165–186
48. Zancani M, Peresson C, Patui S, Tubaro F, Vianello A, Macri F (2007) Mitochondrial ferritin distribution among plant organs and its involvement in ascorbate-mediated iron uptake and release. Plant Sci 173:182–189
49. Munne-Bosch S, Alegre L (2002) Interplay between ascorbic acid and lipophilic antioxidant defences in chloroplasts of water-stressed Arabidopsis plants. FEBS Lett 524:145–148
50. Ranieri A, Lencioni L, Schenone G, Soldatini GF (1993) Glutathione-ascorbic acid cycle in pumpkin plants grown under polluted air in open-top chambers. J Plant Physiol 142:286–290
51. Batini P, Ederli L, Pasqualini S, Antonielli M, Valenti V (1995) Effects of ethylenediurea and ozone in detoxificant ascorbate-ascorbate peroxidase system in tobacco plants. Plant Physiol Biochem 33:717–723
52. Ranieri A, Castagna A, Soldatini GF (2000) Differential stimulation of ascorbate peroxidase isoforms by ozone exposure in sunflower plants. J Plant Physiol 156:266–271
53. Wang WX, Vinocur B, Altman A (2003) Plant responses to drought, salinity and extreme temperatures: towards genetic engineering for stress tolerance. Planta 218:1–14

54. Yabuta Y, Motoki T, Yoshimura K, Takeda T, Ishikawa T, Shigeoka S (2002) Thylakoid membrane-bound ascorbate peroxidase is a limiting factor of antioxidative systems under photo-oxidative stress. Plant J 32:915–925

55. Ivanov BN, Sacksteder CA, Kramer DM, Edwards GE (2001) Light-induced ascorbate-dependent electron transport and membrane energization in chloroplasts of bundle sheath cells of the C4 plant maize. Arch Biochem Biophys 385:145–153

56. Smirnoff N (2000) Ascorbic acid: metabolism and functions of a multi-facetted molecule. Curr Opin Plant Biol 3:229–235

57. Talla S, Riazunnisa K, Padmavathi L, Sunil B, Rajsheel P, Raghavendra AS (2011) Ascorbic acid is a key participant during the interactions between chloroplasts and mitochondria to optimize photosynthesis and protect against photoinhibition. J Biosci 36:163–173

58. Zhang JX, Kirkham MB (1996) Enzymatic responses of the ascorbate-glutathione cycle to drought in sorghum and sunflower plants. Plant Sci 113:139–147

59. Takahama U, Oniki T (1997) A peroxidase/phenolics/ascorbate system can scavenge hydrogen peroxide in plant cells. Physiol Plant 101:845–852

60. Yamasaki H, Grace SC (1998) EPR detection of phytophenoxyl radicals stabilized by zinc ions: evidence for the redox coupling of plant phenolics with ascorbate in the H_2O_2-peroxidase system. FEBS Lett 422:377–380

61. Mehlhorn H, Lelandais M, Korth HG, Foyer CH (1996) Ascorbate is the natural substrate for plant peroxidases. FEBS Lett 378:203–206

62. Battke F, Ernst D, Halbach S (2005) Ascorbate promotes emission of mercury vapour from plants. Plant Cell Environ 28:1487–1495

63. Huang GY, Wang YS, Sun CC, Dong JD, Sun ZX (2010) The effect of multiple heavy metals on ascorbate, glutathione and related enzymes in two mangrove plant seedlings (*Kandelia candel* and *Bruguiera gymnorrhiza*). Oceanol Hydrobiol Stud 39:11–25

64. Fridovich I (1998) Oxygen toxicity: a radical explanation. J Exp Biol 201:1203–1209

65. Asada K (1999) The water–water cycle in chloroplasts: Scavenging of active oxygens and dissipation of excess photons. Ann Rev Plant Physiol Plant Mol Biol 50:601–639

66. Zheng X, Vanhuystee RB (1992) Anionic peroxidase catalyzed ascorbic acid and IAA oxidation in the presence of hydrogen peroxide: a defense system against peroxidative stress in peanut plant. Phytochemistry 31:1895–1898

67. Pignocchi C, Foyer CH (2003) Apoplastic ascorbate metabolism and its role in the regulation of cell signalling. Curr Opin Plant Biol 6:379–389

68. Sanmartin M, Drogoudi PD, Lyons T, Pateraki I, Barnes J, Kanellis AK (2003) Over-expression of ascorbate oxidase in the apoplast of transgenic tobacco results in altered ascorbate and glutathione redox states and increased sensitivity to ozone. Planta 216:918–928

69. Fotopoulos V, De Tullio MC, Barnes J, Kanellis AK (2008) Altered stomatal dynamics in ascorbate oxidase over-expressing tobacco plants suggest a role for dehydroascorbate signalling. J Exp Bot 59:729–737

70. Parsons HT, Fry SC (2010) Reactive oxygen species-induced release of intracellular ascorbate in plant cell-suspension cultures and evidence for pulsing of net release rate. New Phytol 187:332–342

71. Lee EH (1991) Plant resistance mechanisms to air pollutants: rhythms in ascorbic acid production during growth under ozone stress. Chronobiol Int 8:93–102

72. Al-Hakimi AMA, Hamada AM (2001) Counteraction of salinity stress on wheat plants by grain soaking in ascorbic acid, thiamin or sodium salicylate. Biol Plant 44:253–261

73. Teklemariam TA, Sparks JP (2006) Leaf fluxes of NO and NO_2 in four herbaceous plant species: the role of ascorbic acid. Atmos Environ 40:2235–2244

74. Michael PI, Krishnaswamy M (2011) The effect of zinc stress combined with high irradiance stress on membrane damage and antioxidative response in bean seedlings. Environ Exp Bot 74:171–177

75. Younis ME, Hasaneen MNA, Kazamel AMS (2009) Plant growth, metabolism and adaptation in relation to stress conditions. XXVII. Can ascorbic acid modify the adverse effects of NaCl and mannitol on amino acids, nucleic acids and protein patterns in *Vicia faba* seedlings? Protoplasma 235:37–47

76. Hemavathi, Upadhyaya CP, Young KE, Akula N, Kim HS, Heung JJ, Oh OM, Aswath CR, Chun SC, Kim DH, Park SW (2009). Over-expression of strawberry D-galacturonic acid reductase in potato leads to accumulation of vitamin C with enhanced abiotic stress tolerance. Plant Sci 177:659–667

77. Hemavathi, Upadhyaya, CP, Akula N, Young KE, Chun SC, Kim DH, Park SW (2010). Enhanced ascorbic acid accumulation in transgenic potato confers tolerance to various abiotic stresses. Biotechnol Lett 32:321–330

78. Chen Z, Gallie DR (2005) Increasing tolerance to ozone by elevating foliar ascorbic acid confers greater protection against ozone than increasing avoidance. Plant Physiol 138: 1673–1689

79. Kwon SY, Choi SM, Ahn YO, Lee HS, Lee HB, Park YM, Kwak SS (2003) Enhanced stress-tolerance of transgenic tobacco plants expressing a human dehydroascorbate reductase gene. J Plant Physiol 160:347–353

80. Ushimaru T, Nakagawa T, Fujioka Y, Daicho K, Naito M, Yamauchi Y, Nonaka H, Amako K, Yamawaki K, Murata N (2006) Transgenic Arabidopsis plants expressing the rice dehydroascorbate reductase gene are resistant to salt stress. J Plant Physiol 163:1179–1184

81. Eltayeb AE, Kawano N, Badawi GH, Kaminaka H, Sanekata T, Morishima I, Shibahara T, Inanaga S, Tanaka K (2006) Enhanced tolerance to ozone and drought stresses in transgenic tobacco overexpressing dehydroascorbate reductase in cytosol. Physiol Plant 127:57–65

82. Yin LN, Wang SW, Eltayeb AE, Uddin MI, Yamamoto Y, Tsuji W, Takeuchi Y, Tanaka K (2010) Overexpression of dehydroascorbate reductase, but not monodehydroascorbate reductase, confers tolerance to aluminum stress in transgenic tobacco. Planta 231:609–621

83. Adak S, Datta AK (2005) Leishmania major encodes an unusual peroxidase that is a close homologue of plant ascorbate peroxidase: a novel role of the transmembrane domain. Biochem J 390:465–474

84. Hossain Z, Lopez-Climent MF, Arbona V, Perez-Clemente RM, Gomez-Cadenas A (2009) Modulation of the antioxidant system in citrus under waterlogging and subsequent drainage. J Plant Physiol 166:1391–1404

85. Jimenez A, Hernandez JA, delRio LA, Sevilla F (1997) Evidence for the presence of the ascorbate-glutathione cycle in mitochondria and peroxisomes of pea leaves. Plant Physiol 114:275–284

86. Mittler R, Zilinskas BA (1992) Molecular-cloning and characterization of a gene encoding pea cytosolic ascorbate peroxidase. J Biol Chem 267:21802–21807

87. Yoshimura K, Yabuta Y, Ishikawa T, Shigeoka S (2000) Expression of spinach ascorbate peroxidase isoenzymes in response to oxidative stresses. Plant Physiol 123:223–233

88. Agrawal GK, Jwa NS, Iwahashi H, Rakwal R (2003) Importance of ascorbate peroxidases OsAPX1 and OsAPX2 in the rice pathogen response pathways and growth and reproduction revealed by their transcriptional profiling. Gene 322:93–103

89. Wang YJ, Wisniewski M, Meilan R, Cui MG, Webb R, Fuchigami L (2005) Overexpression of cytosolic ascorbate peroxidase in tomato confers tolerance to chilling and salt stress. J Am Soc Hortic Sci 130:167–173

90. Mehlhorn H (1990) Ethylene-promoted ascorbate peroxidase-activity protects plants against hydrogen-peroxide, ozone and paraquat. Plant Cell Environ 13:971–976

91. Lopez F, Vansuyt G, CasseDelbart F, Fourcroy P (1996) Ascorbate peroxidase activity, not the mRNA level, is enhanced in salt-stressed *Raphanus sativus* plants. Physiol Plant 97: 13–20

92. Al-Ghamdi AA (2009) Evaluation of oxidative stress tolerance in two wheat (*Triticum aestivum*) cultivars in response to drought. Int J Agric Biol 11:7–12

93. de Pinto MC, Lavermicocca P, Evidente A, Corsaro MM, Lazzaroni S, De Gara L (2003) Exopolysaccharides produced by plant pathogenic bacteria affect ascorbate metabolism in *Nicotiana tabacum.* Plant Cell Physiol 44:803–810

94. Wang SD, Zhu F, Yuan S, Yang H, Xu F, Shang J, Xu MY, Jia SD, Zhang ZW, Wang JH, Xi DH, Lin HH (2011) The roles of ascorbic acid and glutathione in symptom alleviation to SA-deficient plants infected with RNA viruses. Planta 234:171–181

95. Goggin FL, Avila CA, Lorence A (2010) Vitamin C content in plants is modified by insects and influences susceptibility to herbivory. BioEssays 32:777–790

Chapter 3
Ascorbate Biosynthesis in Plants

The structure of the familiar antioxidant ascorbate was described in 1933, yet its biosynthesis in plants remained elusive in the next half decade. It became clear from radioisotopic labeling studies in the 1950s that plant ascorbate biosynthesis does not proceed in a route similar to that in mammals. Although ascorbate biosynthesis and metabolism pathway in animals has been elucidated in 1960s, the counterpart pathway in plants was not identified until the Smirnoff pathway was proposed in the late 1990s. The description in 1996 of an Arabidopsis mutant deficient in ascorbate prompted renewed research effort in this area, and subsequently in 1998, D-mannitol/L-galactose pathway was proposed that is backed by strong biochemical and molecular genetic evidence. This pathway proceeds through the intermediates GDP-D-mannose, L-galactose, and L-galactono-1, 4-lactone [1, 2].

After then, several other pathways have been successively put forward known as galacturonate pathway, gulose pathway and myoinositol pathway. The last step enzyme in gulose pathway and myoinositol pathway in plants is overlapped with that in animals, thus combining together the ascorbate biosynthesis pathways in animals and plants. GDP-L-gulose and myoinositol are proposed as new intermediates in ascorbate biosynthesis, indicating that part of the animal pathway might also be operating in plants (Fig. 3.1). Unlike the unique pathway for ascorbate biosynthesis and metabolism in animals, several alternative pathways for ascorbate biosynthesis and metabolism in plants make their ascorbate synthesis more complicated.

An elucidation of the biosynthesis of ascorbate as well as the study of alternative proposed pathways should broaden our understanding of ascorbate metabolism in plants. The EMS-induced Arabidopsis mutants screened by sensitivity to ozone and decreasing ascorbate accumulation have paved the way for elucidating the ascorbate biosynthesis pathway [3]. On the other hand, as the competence for ascorbate synthesis seems to be ubiquitous amongst plant cells [4], whether there is any other alternative biosynthesis pathway in plants is intriguing people's interest. The genome-wide association analysis has shown its potency in plant genetics, which might help to decipher the complexity of ascorbate biosynthesis.

Y. Zhang, *Ascorbic Acid in Plants*, SpringerBriefs in Plant Science,
DOI: 10.1007/978-1-4614-4127-4_3, © The Author 2013

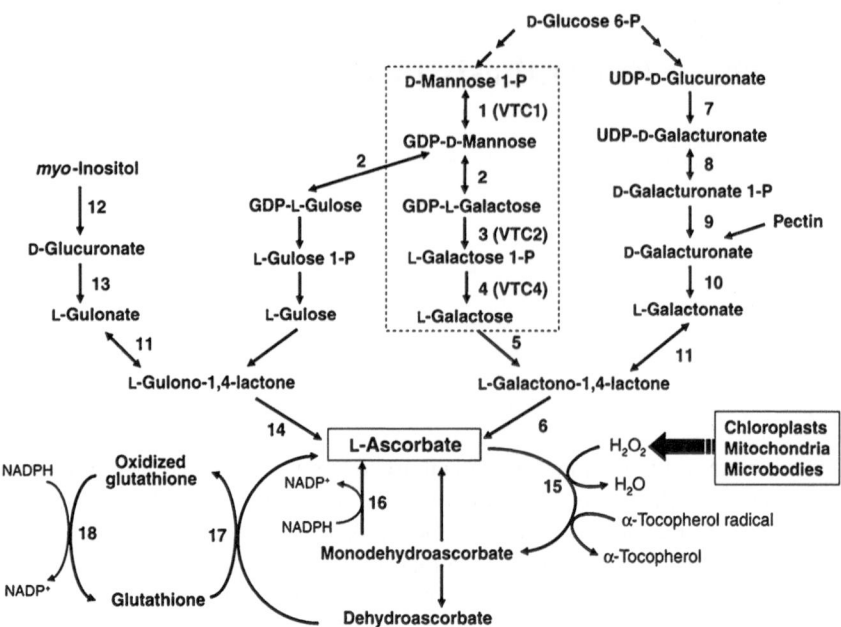

Fig. 3.1 Proposed ascorbate biosynthetic pathways and ascorbate–glutathione cycle in plants. The broken line box shows D-mannose/L-galactose pathway. Parentheses indicate the names of ascorbate-deficient (VTC) Arabidopsis mutants. Enzymes: (*1*) GDP-mannose pyrophosphorylase; (*2*) GDP-mannose -3′,5′-epimerase; (*3*) GDP-L-galactose phosphorylase/L- galactose guanylyl-transferase; (*4*) L-galactose 1-phosphate phosphatase; (*5*) L-galactose dehydrogenase; (*6*) L-galactono-1,4-lactone dehydrogenase; (*7*) UDP-glucuronate 4-epimerase; (*8*) UDP-galacturonate pyrophosphatase or phosphorylase or uridylyltransferase; (*9*) D-galacturonate 1-phosphate phosphatase; (*10*) D-galacturonate reductase; (*11*) aldonolactonase; (*12*) *myo*inositol oxygenase; (*13*) D-glucuronate reductase; (*14*) L-GulL dehydrogenase or oxidase; (*15*) ascorbate peroxidase; (*16*) monodehydroascorbate reductase; (*17*) dehydroascorbate reductase; (*18*) glutathione reductase

3.1 D-Man/L-Gal Pathway

Biochemical and molecular genetic evidence supports synthesis from GDP-D-mannose via L-galactose (D-Man/L-Gal pathway) as a significant source of ascorbate. The early isotopic tracing showed that isotope-tagged D-mannose was transformed into ascorbate through the intermediate product of L-galactose, thus demonstrating the D-mannose and L-galactose are important precursors for ascorbate biosynthesis [1]. Other works showed that exogenous L-galactose and L-galactono-1,4-lactone can be quickly oxidized into ascorbate under the catalysis of GalDH and GLDH [1].

After the identification of an enzyme in pea and Arabidopsis, L-galactose dehydrogenase, that catalyses oxidation of L-galactose to L-galactono-1,4-lactone, Wheeler et al. proposed an ascorbate biosynthesis pathway involving GDP-

D-mannose, GDP-L-galactose, L-galactose and L-galactono-1,4-lactone, and have synthesized ascorbate from GDP-D-mannose utilizing these intermediates in vitro [1]. Based on the plenty of previous results, Smirnoff et al. proposed the ascorbate biosynthesis pathway of D-Man/L-Gal, also known as Smirnoff pathway.

Following the proposing of D-Man/L-Gal pathway, numerous reports demonstrate its rationality. Antisense inhibition of *GalDH* in Arabidopsis [5] and *GLDH* in tobacco [3] both lead to reduced ascorbate content in plants, demonstrating the *GalDH* and *GLDH* play important roles in ascorbate biosynthesis. Conklin et al. screened the Arabidopsis defective mutants in ascorbate, and identified 5 loci related to ascorbate biosynthesis: *VTC1*, *VTC2*, *VTC3*, *VTC4* and *VTC5*. All those loci except *VTC3* have been cloned and confirmed as important genes responsible for ascorbate biosynthesis. Among those loci, *VTC1* encodes a GDP-mannose pyro-phosphorylase (GMP), *VTC2* and *VTC5* both encode GDP-L-galactose phosphory-lase (GGP), and *VTC4* encodes L- galactose 1-phosphate phosphatase (GPP) [6]. The double mutant generated by crossing *vtc2* and *vtc5* exhibited growth inhibition after seed germination with whitish cotyledon. The growth retardation can be removed by subsequent addition of ascorbate or precursor of L-galactose [7]. Overexpressing of another important enzyme in D-Man/L-Gal pathway, GME, leads to ascorbate accumulation with enhanced tolerance to oxidative stress, cold, and salt in tomato [8]. All these findings demonstrate the importance of D-Man/L-Gal pathway for ascorbate biosynthesis.

In D-Man/L-Gal pathway, ascorbate was synthesized from the precursor of D-glucose through more than ten enzymatic reaction steps. The upstream six steps are responsible for biosynthesis of nucleotide sugar, which serves as the substrate for ascorbate biosynthesis as well as precursor for cell wall polysaccharides and glycoproteins. The last four enzymatic steps starting from GDP-L-galactose are unique for ascorbate biosynthesis. GGP is the first step enzyme in this unique pathway. Up to date, all the genes encoding the necessary enzymes in the D-Man/L-Gal pathway have been cloned from Arabidopsis. This pathway is considered the main ascorbate biosynthesis machinery in plants, and has been demonstrated in various plant species. However, other possible biosynthesis pathways for ascorbate can not be excluded in plants.

3.2 Myoinositol Pathway

Myoinositol pathway for ascorbate biosynthesis in plants is composed of several enzymatic steps. The glucose 6-phosphate derivative, myoinositol-1-phosphate, is transformed into myoinositol under the catalytic reaction of phosphatase [9], and then the myoinositol is oxidized into D-glucuronate. D-glucuronate is then converted into L-gulonic acid under the catalysis of glucuronate reductase, and the L-gulonic acid is transformed into L-gulono-1,4-lactone by aldonolactonase. The L-gulono-1,4-lactone is finally converted into ascorbate by L-gulonolactone dehy-drogenase/oxidase [10].

Although the myoinositol is not the main precursor for ascorbate biosynthesis, current research results indicate regulating myoinositol metabolism could promote the accumulation of ascorbate in plants, which also serves as genetic evidence for myoinositol pathway. Molecular and biochemical evidence supports a possible biosynthetic route using myoinositol as the initial substrate. A myoinositol oxygenase (*MIOX*) gene was identified in chromosome 4 (*miox4*) of Arabidopsis, and its enzymatic activity was confirmed in bacterially expressed recombinant protein. *miox4* was primarily expressed in flowers and leaves of wild-type Arabidopsis plants, tissues with a high concentration of ascorbate. Overexpressing *miox4* open reading frame leads to 2- to 3-fold increasing in the ascorbate content in Arabidopsis leaves, suggesting the role of myoinositol in ascorbate biosynthesis and the potential of using *miox4* gene for the agronomic and nutritional enhancement of crops [11].

Another report indicates that overexpressing a gene encoding purple acid phosphotase (*AtPAP15*), involved in myoinositol metabolism, could promote the ascorbate accumulation to 2 folds of the wild-type [12]. In our previous work, overexpression of strawberry *MIOX* gene in tomato (cv. Ailsa Craig) could not improve the ascorbate content in tomato leaves significantly. However, when fed with myoinositol, the leaf ascorbate content in *MIOX* transgenic tomato plants improved significantly while the wild-type plants stayed with the same amount. We suggest that myoinositol is involved in ascorbate biosynthesis and thus give evidence of the myoinositol pathway in tomato. On the other hand, the endogenous myoinositol amount in plant is probably insufficient to activate the myoinositol pathway, even if the *MIOX* gene expression is strengthened (unpublished data). All these findings indicate the myoinositol pathway could be an alternative mechanism for ascorbate biosynthesis.

3.3 Galacturonate Pathway

Early in 1960s, isotopic tracing showed galacturonic acid methyl ester could be transformed into ascorbate without the reaction of hexose monophosphate pathway. In 1999, Davey et al. found that galacturonic acid methyl ester could be effectively transformed into ascorbate in Arabidopsis cell suspension culture [13]. The researchers thus predict that besides the D-Man/L-Gal pathway, plants possess an alternative ascorbate biosynthesis pathway in which galacturonic acid is involved. Until 2003, Agius et al. cloned a gene encoding NADPH dependent D-galacturonate reductase (GalUR, EC 1.1.1.19) from strawberry, and proposed the galacturonate pathway for ascorbate biosynthesis in strawberry fruit [13]. The research results indicated the close correlation between *GalUR* gene expression and ascorbate accumulation along with fruit development of strawberry [13]. The substrate of galacturonate reductase, galacturonic acid, is one of the major components of cell wall pectin, and the antisense inhibition of pectate lyase could reduce the ascorbate accumulation in strawberry fruit. The subsequent overexpression of the gene encoding strawberry galacturonate reductase in Arabidopsis could significantly improve the ascorbate content (2–3 folds) as compared to control [14]. Also, overexpression of strawberry galacturonate reductase in tomato could improve fruit ascorbate accumulation. The overexpression of *GalUR* could up-regulate the expression level of L-galactono-1,4-lactone dehydrogenase gene (*GLDH*), the last step enzyme encoding gene in D-Man/L-Gal pathway, which indicates the importance of galacturonate pathway in ascorbate biosynthesis and the interaction of the biosynthesis pathways (unpublished data).

The currently proposed galacturonate pathway is composed of several key enzymatic steps. The D-galacturonic acid is reduced to L- galactonic acid or L-galactono-1,4-lactone by galacturonate reductase, and then the reduction product is transformed into ascorbate by GLDH or other key enzymes.

More and more evidence support the existence of galacturonate pathway for plant ascorbate biosynthesis. When fed with D-galacturonic acid through the hairy roots, the ascorbate content in transgenic tomato overexpressing strawberry *GalUR* gene as well as untransformed control plants was both improved, while the ascorbate content in hairy roots of transgenic plants was more significantly enhanced (up to 2.5 folds) as compared the control plants fed with water [14]. These findings demonstrated that galacturonate pathway existed in hairy roots of tomato. Overexpressing the strawberry *GalUR* in potato, can promote the ascorbate accumulation in potato leaves and tubers [15]. Di Matteo found that ascorbate in fruits of tomato introgression line IL12-4 was higher than that of parent plant, M82, among several trial seasons. Further investigation indicated that fruit of IL12-4 accumulates more D-galacturonic acid by pectin degrading, thus leading to accumulation of ascorbate through galacturonate pathway in tomato [16]. However, it remains to be investigated that whether the galacturonate pathway performs generally in tomato or only happens in introgression line of IL12-4.

On the other hand, the overexpression of strawberry *GalUR* gene in tomato improved the *GLDH* gene expression, and increased the ascorbate accumulation in tomato fruit (unpublished data). The galacturonate pathway is also proposed in grape via analyzing genes involved in ascorbate biosynthesis during growth and ripening of grape [17]. Although several findings suggested the rationality of galacturonate pathway in plants, *GalUR* was the only gene identified up to date in this pathway, and thus more work and evidence are needed for further characterizing the pathway.

3.4 Gulose Pathway

When studying the biochemistry features of GME in Arabidopsis, Wolucka et al. found that besides the 3',5-isomerization activity of converting GDP-D-mannose to GDP-L-galactose, GME also possess the 5' isomerization activity of catalyzing GDP-D-mannose into GDP-L-gulose [9]. GDP-L-galactose could act as the as donor of L-galactose residue in the biosynthesis of cell polysaccharide and glycoprotein.

However, similar function was not found thus far for L-gulose residue in plant cells. It is suggested that GDP-L-gulose mainly functions as substrate in ascorbate biosynthesis. The L-gulose derivative GDP-L-gulose is likely oxidized by L-galactose dehydrogenase to form L-gulono-1,4-lactone. L-gulono-1,4-lactone is not the substrate of L-galactono-1,4-lactone dehydrogenase but the substrate of L-gulono-1,4-lactone dehydrogenase, which catalyzes L-gulono-1,4-lactone into ascorbate. The enzyme activity of L-gulono-1,4-lactone dehydrogenase has been detected in mitochondria of potato [18].

Jain and Nessler overexpressed the gene encoding L-gulono-1,4-lactone dehydrogenase from rat in plant cells, and the ascorbate content in transgenic plants was improved significantly [19]. Overexpression of rat L-gulono-1,4-lactone dehydrogenase gene in ascorbate defective mutant *vtc1* could recover the ascorbate content as high as wild-type plants [20]. Seven genes with homology with rat L-gulono-1,4-lactone dehydrogenase gene were found in Arabidopsis genome. Maruta et al. overexpressed the 7 genes respectively in tobacco BY2 cell suspension culture, and after the treatment with L-gulono-1,4-lactone, the cell line transformed with *AtGulLO2*, *AtGulLO3* and *AtGulLO5* showed increased ascorbate accumulation [21]. All these data indicate gulose pathway is the alternative machinery for ascorbate biosynthesis, and the L-gulono-1,4-lactone dehydrogenase is the potential rate-determining step for the gulose pathway. More evidence is needed to support the gulose pathway for ascorbate biosynthesis in plants.

3.5 VTC2 Cycle

It has been shown that the *VTC2* encoded enzyme possesses the activity of D-mannose-1-P guanylyltransferase, which overlapped with VTC1. Thus, Laing et al. proposed that VTC2 circle for ascorbate synthesis may exist in plant kingdom. Only two enzymes, VTC2 and GDP-mannose-3′, 5′-epimerase, are supposedly involved in the VTC2 circle [22]. Under the catalysis of VTC2 transferase, D-mannose-1-phosphate is transformed into GDP-D-mannose. At the same time, another substrate, GDP-L-galactose, is transformed into L-galactose-1-phosphate. The product of L-galactose-1-phosphate will be utilized for ascorbate biosynthesis, and the other product of GDP-D-mannose will be converted back to GDP-L-galactose under the catalyzing of GME. VTC2 will then catalyze the GDP-L-galactose as substrate and this reaction circle is called VTC2 circle (Fig. 3.2). In this reaction circle, VTC1 (GMP) enzyme is not so much involved as in other ascorbate biosynthesis pathway. Only when GDP-D-mannose is not sufficient for cells polysaccharide and glycoprotein biosynthesis, are the VTC1 needed to catalyze the D-mannose-1-phosphate into GDP-D-mannose, which will then be used for polysaccharide and glycoprotein formation.

The observation that the VTC2 enzyme can use glucose 1-phosphate and GDP-D-glucose as substrates, and the presence of a GDP-D-mannose 2′-epimerase activity, have led to the proposal of an extended VTC2 cycle that links

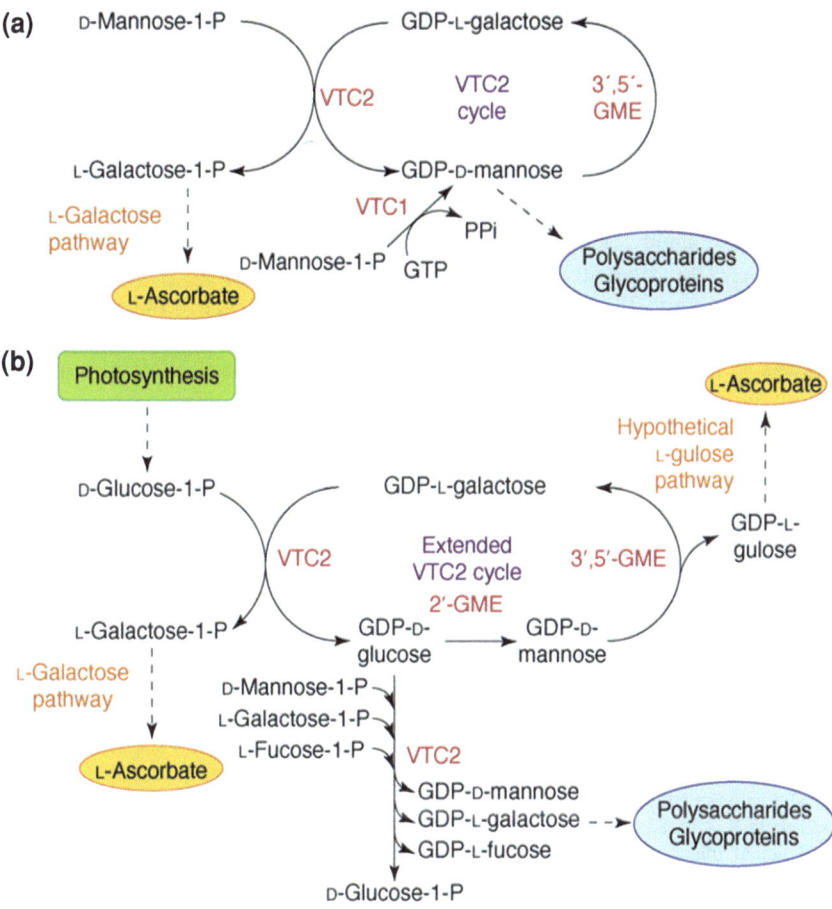

Fig. 3.2 **a** The VTC2 circle proposed by Laing et al. **b** The extended VTC2 circle proposed by Wolucka et al. (cited from [24])

photosynthesis with the biosynthesis of ascorbate and the cell wall metabolism in plants. Wolucka and Van Montagu then improved the VTC2 circle [23]. The improved VTC2 circle takes it into consideration that hexosemonophosphate can be utilized as receptor of guanylyl, and GME also has the isomerization activity of transforming GDP-D-glucose into GDP-D-mannose. In the improved VTC2 circle, the photosynthate of D-glucose-1-phosphate can participate in the ascorbate biosynthesis directly without the catalyzing of GME enzyme. In the VTC2 circle, L-galactose-1-phosphate can be used for ascorbate biosynthesis, and other product of GDP-hexose can be involved in biosynthesis of cell polysaccharides and glycoprotein, with energy being kept in GDP-hexose in form of phosphodiester bond. Thus the guanylyltransferase activity is critical for the VTC2 cycle and the extended VTC2 cycle.

However, researchers' debates make the VTC2 circle controversial. The main criticism include: Firstly, transferase activity of VTC2 was not validated in other investigation [23]; Secondly, double isotope labeling test indicated that GDP-D-mannose is synthesized predominantly via a pathway involving PMI rather than through a pathway involving epimerization from GDP-D-glucose to GDP-D-mannose. Therefore, VTC2 circle needs further verification, especially whether GME enzyme has the 2' isomerization activity and whether hexosemonophosphate could act as guanylyl-acceptors for VTC2 enzyme remained to be confirmed.

Evidence for alternative ascorbate biosynthetic pathways exist and await additional scrutiny. The increasing Arabidopsis mutants deficient in ascorbate will further increase our understanding of ascorbate biosynthesis. With a biosynthetic pathway in hand, research on areas such as the control of ascorbate biosynthesis and the physiological roles of ascorbate should progress rapidly.

References

1. Wheeler GL, Jones MA, Smirnoff N (1998) The biosynthetic pathway of vitamin C in higher plants. Nature 393:365–369
2. Smirnoff N, Conklin PL, Loewus FA (2001) Biosynthesis of ascorbic acid in plants: a renaissance. Annu Rev Plant Biol 52:437–467
3. Conklin PL, Norris SR, Wheeler GL, Williams EH, Smirnoff N, Last RL (1999) Genetic evidence for the role of GDP-mannose in plant ascorbic acid (vitamin C) biosynthesis. Proc Natl Acad Sci U S A 96:4198–4203
4. De Tullio MC, Arrigoni O (2004) Hopes, disillusions and more hopes from vitamin C. Cell Mol Life Sci 61:209–219
5. Tabata K, Oba K, Suzuki K, Esaka M (2001) Generation and properties of ascorbic acid-deficient transgenic tobacco cells expressing antisense RNA for L-galactono-1,4-lactone dehydrogenase. Plant J 27:139–148
6. Dowdle J, Ishikawa T, Gatzek S, Rolinski S, Smirnoff N (2007) Two genes in *Arabidopsis thaliana* encoding GDP-L-galactose phosphorylase are required for ascorbate biosynthesis and seedling viability. Plant J 52:673–689
7. Zhang CJ, Liu JX, Zhang YY, Cai XF, Gong PJ, Zhang JH, Wang TT, Li HX, Ye ZB (2011) Overexpression of SlGMEs leads to ascorbate accumulation with enhanced oxidative stress, cold, and salt tolerance in tomato. Plant Cell Rep 30:389–398
8. Gillaspy GE, Keddie JS, Oda K, Gruissem W (1995) Plant inositol monophosphatase is a lithium-sensitive enzyme encoded by a multigene family. Plant Cell 7:2175–2185
9. Wolucka BA, Van Montagu M (2003) GDP-mannose 3', 5'-epimerase forms GDP-L-gulose, a putative intermediate for the de novo biosynthesis of vitamin C in plants. J Biol Chem 278:47483–47490
10. Lorence A, Chevone BI, Mendes P, Nessler CL (2004) myo-inositol oxygenase offers a possible entry point into plant ascorbate biosynthesis. Plant Physiol 134:1200–1205
11. Zhang W, Gruszewski HA, Chevone BI, Nessler CL (2008) An Arabidopsis purple acid phosphatase with phytase activity increases foliar ascorbate. Plant Physiol 146:431–440
12. Davey MW, Gilot C, Persiau G, Ostergaard J, Han Y, Bauw GC, Van Montagu MC (1999) Ascorbate biosynthesis in Arabidopsis cell suspension culture. Plant Physiol 121:535–543
13. Agius F, Gonzalez-Lamothe R, Caballero JL, Munoz-Blanco J, Botella MA, Valpuesta V (2003) Engineering increased vitamin C levels in plants by overexpression of a D-galacturonic acid reductase. Nat Biotechnol 21:177–181

14. Oller ALW, Agostini E, Milrad SR, Medina MI (2009) In situ and de novo biosynthesis of vitamin C in wild type and transgenic tomato hairy roots: a precursor feeding study. Plant Sci 177:28–34

15. Hemavathi, Upadhyaya CP, Young KE, Akula N, Kim HS, Heung JJ, Oh OM, Aswath CR, Chun SC, Kim DH, Park SW (2009) Over-expression of strawberry D-galacturonic acid reductase in potato leads to accumulation of vitamin C with enhanced abiotic stress tolerance. Plant Sci 177:659–667

16. Di Matteo A, Sacco A, Anacleria M, Pezzotti M, Delledonne M, Ferrarini A, Frusciante L, Barone A (2010) The ascorbic acid content of tomato fruits is associated with the expression of genes involved in pectin degradation. BMC Plant Biol 10:163

17. Cruz-Rus E, Botella MA, Valpuesta V, Gomez-Jimenez MC (2010) Analysis of genes involved in L-ascorbic acid biosynthesis during growth and ripening of grape berries. J Plant Physiol 167:739–748

18. Jain AK, Nessler CL (2000) Metabolic engineering of an alternative pathway for ascorbic acid biosynthesis in plants. Mol Breed 6:73–78

19. Radzio JA, Lorence A, Chevone BI, Nessler CL (2003) L-Gulono-1,4-lactone oxidase expression rescues vitamin C-deficient Arabidopsis (vtc) mutants. Plant Mol Biol 53:837–844

20. Maruta T, Ichikawa Y, Mieda T, Takeda T, Tamoi M, Yabuta Y, Ishikawa T, Shigeoka S (2010) The contribution of Arabidopsis homologs of 1-gulono-1,4-lactone oxidase to the biosynthesis of ascorbic acid. Biosci Biotechnol Biochem 74:1494–1497

21. Laing WA, Wright MA, Cooney J, Bulley SM (2007) The missing step of the L-galactose pathway of ascorbate biosynthesis in plants, an L-galactose guanyltransferase, increases leaf ascorbate content. Proc Natl Acad Sci U S A 104:9534–9539

22. Wolucka BA, Van Montagu M (2007) The VTC2 cycle and the de novo biosynthesis pathways for vitamin C in plants: an opinion. Phytochemistry 68:2602–2613

23. Linster CL, Adler LN, Webb K, Christensen KC, Brenner C, Clarke SG (2008) A second GDP-L-galactose phosphorylase in Arabidopsis en route to vitamin C-covalent intermediate and substrate requirements for the conserved reaction. J Biol Chem 283:18483–18492

24. Linster CL, Clarke SG (2008) L-Ascorbate biosynthesis in higher plants: the role of VTC2. Trends Plant Sci 13:567–573

Chapter 4
The Oxidization and Catabolism of Ascorbate

Recent progress has revealed biosynthetic pathways for ascorbate, but the degradative pathways remain largely elusive. Ascorbate catabolism involves reversible oxidation to dehydroascorbate, followed by irreversible oxidation or hydrolysis. The precursor-product relationships and the identity of several major dehydroascorbate breakdown products remain undetermined.

Ascorbate acts in the cell metabolism as an electron donor, and consequently ascorbate free radical (monodehydroascorbate) is continuously produced. Monodehydroascorbate can be reconverted to ascorbate by means of monodehydroascorbate reductase or can undergo spontaneous disproportion, generating dehydroascorbate [1].

The ascorbate is biosynthesized in plant cells as the reduced status of ascorbate, and can then be oxidized by ascorbate oxidase (AO) or ascorbate peroxidase (APX). AO is a multi-copper enzyme located in apoplast. AO catalyzes the oxidation of ascorbate to dehydroascorbate. At the same time, AO reduces molecular oxygen into water. AO is transcribed at high abundance in fast-growing tissues, e.g. expanding leaves, elongating stem, growing roots and swelling melon crops [2]. These indicate that AO may play pivotal roles in cell growth and expansion [3, 4]. *AO* genes exist usually in form of gene family. At least four *AO* genes were found in melon [5]. The growth rate of tobacco plant was promoted after overexpressing *AO* gene in tobacco, while the total ascorbate content in tobacco leaves remained unchanged [6]. Overexpressing the melon *AO* gene in tobacco, however, led to the oxidization of ascorbate in leaf apoplast, altered redox state of ascorbate and glutathione in plant cells, and increased sensitivity to ozone in transgenic plants [7]. On the other hand, suppressing the *AO* gene expression in tobacco, reduced the ascorbate oxidization and the ratio of dehydroascorbate/ascorbate in apoplast, and improved the salt tolerance of transgenic tobacco plants [8]. These results indicate that AO enzyme activity influences the ascorbate redox state, and alters plants stress tolerance since ascorbate acts as an important antioxidant.

APX oxidizes the ascorbate and at same time reduces the hydrogen peroxide into water. In this redox process, the surplus ROS in plant cells can be scavenged. APX

Y. Zhang, *Ascorbic Acid in Plants*, SpringerBriefs in Plant Science, DOI: 10.1007/978-1-4614-4127-4_4, © The Author 2013

exhibits high specificity upon its substrate ascorbate and hydrogen peroxide. Just like ascorbate oxidase gene, APX also belongs to a multi-gene family. APX gene family is composed of several homologous genes in various compartments of plant cells, e.g. cytoplasm, chloroplast, mitochondria and peroxisome [9]. Several *APX* genes in different compartments of Arabidopsis, rice and tomato have been characterized. At least six *APX* genes were found in Arabidopsis, two were located in cytoplasm, two were in peroxisome, one in mitochondria, and one in chloroplast [10]. Rice has at least eight *APX* genes, two are located in cytoplasm, two in peroxisome, three in chloroplast, and one in mitochondria [11]. And in tomato, seven *APX* gene have been identified, three of them are located in cytoplasm, two in peroxisome, and two in chloroplast [12]. Phylogenetic tree and gene structure analysis indicate that different *APX* genes might come up as an result of complicated gene evolution process like gene overlapping [13]. Because the *APX* gene is very important in scavenging the ROS, *APX* is frequently utilized in abiotic stress tolerance in plants. When utilizing APX for engineering abiotic stress tolerance in plants, two different ways of gene overexpression and suppression are used. Both approaches are reported to improve the plant abiotic stress tolerance [14, 15]. The overexpressing approach can improve the APX enzyme activity, which is beneficial for scavenging of ROS. On the other hand, the suppression approach can reduce the oxidation of ascorbate by suppressing APX, thus improve the ascorbate accumulation, which is an important antioxidant.

Dehydroascorbate is shown to be catabolized to at least three products such as oxalate, L-threonate and L-tartrate. The enzymes involved have not been identified, so catabolism is not yet amenable to manipulation [16]. In some plants (*Vitaceae*), ascorbate is degraded via L-idonate to L-tartrate, with the latter arising from carbons 1-4 of ascorbate. The rate-limiting step of the direct pathway from ascorbate to tartaric acid in higher plants is proposed [17]. In most plants, however (including *Vitaceae*), ascorbate degradation can occur via dehydroascorbate, yielding oxalate plus L-threonate, with the latter from carbons 3-6 of ascorbate. The metabolic pathway operates extracellularly in cultured Rosa cells, proceeds via several novel intermediates. Two novel compounds have been detected as apoplastic intermediates in the ascorbate degradation: namely, 4-O-oxalyl-L-threonate and cyclic oxalyl di-ester of L-threonate. Parsons et al. [18] found that the intracellular [^{14}C]ascorbate is irreversibly catabolized to [^{14}C]oxalyl-threonate and related products in the late-growth-phase cultured cells. Two novel oxalyl-esterase activities are also shown to be involved in the ascorbate degradation pathway [19].

The ascorbate metabolic pathway can also operate non-enzymatically, potentially accounting for vitamin losses during cooking. Several steps in the pathway may generate peroxide; this may contribute to the role of ascorbate as a prooxidant besides its better-known capacity as an antioxidant. Acting as a prooxidant in addition to antioxidant, apoplastic ascorbate may loosen the cell wall and hence promote cell expansion and/or fruit softening [20].

It is proposed that dehydroascorbate is a branch-point in ascorbate catabolism, being either oxidized to oxalate and its esters or hydrolysed to 2,3-dioxo-L-gulonate and downstream carboxypentonates. The oxidation/hydrolysis ratio is governed by

ROS status. In vivo, oxalyl esters are enzymatically hydrolysed, but the carboxy-pentonates are stable [21].

In the presence of added hydrogen peroxide, dehydroascorbate undergoes little hydrolysis to 2,3-dioxo-L-gulonate. Instead, it yields oxalyl L-threonate, cyclic oxalyl L-threonate and free oxalate (~ 6:1:1), essentially simultaneously, suggesting that all three product classes independently arose from one reactive intermediate, proposed to be cyclic-2,3-O-oxalyl-L-threonolactone. Only with plant apoplastic esterases present were the esters significant precursors of free oxalate.

Without added hydrogen peroxide, dehydroascorbate was slowly hydrolysed to 2,3-dioxo-L-gulonate. Downstream of 2,3-dioxo-L-gulonate is a singly ionized dicarboxy compound (suggested to be 2-carboxy-L-xylonolactone plus 2-carboxy-L-lyxonolactone), which is reversibly de-lactonized to a dianionic carboxypento-nate. Formation of these lactones and acid was minimized by the presence of residual unreacted ascorbate. In vivo, the putative 2-carboxy-L-pentonolactones are relatively stable [21].

Recent developments in ascorbate metabolism demonstrate its function as a precursor for specific processes in the biosynthesis of organic acids. Ascorbate catabolism and the formation of oxalic and L-tartaric acids provide compelling evidence for a major role of ascorbate in plant metabolism. Combined experimental approaches, using classic biochemical and emerging 'omics' technologies, may provide insight to its metabolism mechanism [22].

References

1. Arrigoni O (1994) Ascorbate system in plant development. J Bioenerg Biomembr 26:407–419
2. Kato N, Esaka M (1996) cDNA cloning and gene expression of ascorbate oxidase in tobacco. Plant Mol Biol 30:833–837
3. Kato N, Esaka M (1999) Changes in ascorbate oxidase gene expression and ascorbate levels in cell division and cell elongation in tobacco cells. Physiol Plant 105:321–329
4. Kato N, Esaka M (2000) Expansion of transgenic tobacco protoplasts expressing pumpkin ascorbate oxidase is more rapid than that of wild-type protoplasts. Planta 210:1018–1022
5. Sanmartin M, Pateraki I, Chatzopoulou F, Kanellis AK (2007) Differential expression of the ascorbate oxidase multigene family during fruit development and in response to stress. Planta 225:873–885
6. Pignocchi C, Fletcher JM, Wilkinson JE, Barnes JD, Foyer CH (2003) The function of ascorbate oxidase in tobacco. Plant Physiol 132:1631–1641
7. Sanmartin M, Drogoudi PD, Lyons T, Pateraki I, Barnes J, Kanellis AK (2003) Over-expression of ascorbate oxidase in the apoplast of transgenic tobacco results in altered ascorbate and glutathione redox states and increased sensitivity to ozone. Planta 216:918–928
8. Yamamoto A, Bhuiyan NH, Waditee R, Tanaka Y, Esaka M, Oba K, Jagendorf AT, Takabe T (2005) Suppressed expression of the apoplastic ascorbate oxidase gene increases salt tolerance in tobacco and Arabidopsis plants. J Exp Bot 56:1785–1796

9. Jimenez A, Hernandez JA, delRio LA, Sevilla F (1997) Evidence for the presence of the ascorbate-glutathione cycle in mitochondria and peroxisomes of pea leaves. Plant Physiol 114:275–284

10. Chew O, Whelan J, Millar AH (2003) Molecular definition of the ascorbate-glutathione cycle in Arabidopsis mitochondria reveals dual targeting of antioxidant defenses in plants. J Biol Chem 278:46869–46877

11. Teixeira FK, Menezes-Benavente L, Galvao VC, Margis R, Margis-Pinheiro M (2006) Rice ascorbate peroxidase gene family encodes functionally diverse isoforms localized in different subcellular compartments. Planta 224:300–314

12. Najami N, Janda T, Barriah W, Kayam G, Tal M, Guy M, Volokita M (2008) Ascorbate peroxidase gene family in tomato: its identification and characterization. Mol Genet Genomics 279:171–182

13. Teixeira FK, Menezes-Benavente L, Margis R, Margis-Pinheiro M (2004) Analysis of the molecular evolutionary history of the ascorbate peroxidase gene family: Inferences from the rice genome. J Mol Evol 59:761–770

14. Zhang YY, Li HX, Shu WB, Zhang CJ, Ye ZB (2011a) RNA interference of a mitochondrial *APX* gene improves vitamin C accumulation in tomato fruit. Sci Hortic 129:220–226

15. Zhang CJ, Liu JX, Zhang YY, Cai XF, Gong PJ, Zhang JH, Wang TT, Li HX, Ye ZB (2011b) Overexpression of *SlGMEs* leads to ascorbate accumulation with enhanced oxidative stress, cold, and salt tolerance in tomato. Plant Cell Rep 30:389–398

16. Hancock RD, Viola R (2005) Biosynthesis and catabolism of L-ascorbic acid in plants. Crit Rev Plant Sci 24:167–188

17. DeBolt S, Cook DR, Ford CM (2006) L-Tartaric acid synthesis from vitamin C in higher plants. Proc Nat Acad Sci USA 103:5608–5613

18. Parsons HT, Fry SC (2010) Reactive oxygen species-induced release of intracellular ascorbate in plant cell-suspension cultures and evidence for pulsing of net release rate. New Phytol 187:332–342

19. Green MA, Fry SC (2005a) Apoplastic degradation of ascorbate: novel enzymes and metabolites permeating the plant cell wall. Plant Biosys 139:2–7

20. Green MA, Fry SC (2005b) Vitamin C degradation in plant cells via enzymatic hydrolysis of 4-O-oxalyl-L-threonate. Nature 433:83–87

21. Parsons HT, Yasmin T, Fry SC (2011) Alternative pathways of dehydroascorbic acid degradation in vitro and in plant cell cultures: novel insights into vitamin C catabolism. Biochem J 440:375–383

22. Debolt S, Melino V, Ford CM (2007) Ascorbate as a biosynthetic precursor in plants. Ann Bot Lond 99:3–8

Chapter 5
Recycling of Ascorbate

After the ascorbate is synthesized and metabolically oxidized in plants, part of the metabolic products can be recycled to the reduced state of ascorbate.

The reduced ascorbate can be oxidized into monodehydroascorbate; the monodehydroascorbate is not stable and can be reduced again by MDHAR into the reduced state of ascorbate. Monodehydroascorbate can also be converted into dehydroascorbate via non-enzymatic reaction system. The potentially unstable dehydroascorbate can be degraded directly into 2,3-diketo-L-gulonic acid, or transformed into reduced ascorbate by catalysis of DHAR. As glutathione is needed as reducer in the recycling of ascorbate, the recycling process is called as ascorbate–glutathione circle. The overexpression of glutathione-dependent dehydroascorbate reductase has resulted in increased ascorbate.

Numerous reports showed that it is possible to improve ascorbate accumulation in plant cells via regulating the ascorbate recycling process. Chen et al. overexpressed the *DHAR* gene from wheat in tobacco and maize, and the *DHAR* gene expression level was improved to 32 folds and 100 folds, respectively. The ascorbate content in transgenic tobacco and maize increased 2–4 folds of the wild-type. In transgenic plants, the redox state of ascorbate was enhanced, and the content of glutathione utilized by DHAR for ascorbate recycling was also improved [1]. Eltayeb et al. found that overexpressing *MDHAR* in tobacco not only promoted the accumulation of reduced ascorbate, but also improved the plant resistance to ozone, salt, and drought stress [2]. Similarly, overexpressing of another gene involved in ascorbate recycling, *DHAR*, improved the plant tolerance to ozone and drought stress in transgenic tobacco [3].

References

1. Chen Z, Young TE, Ling J, Chang SC, Gallie DR (2003) Increasing vitamin C content of plants through enhanced ascorbate recycling. Proc Nat Acad Sci USA 100:3525–3530

2. Eltayeb AE, Kawano N, Badawi GH, Kaminaka H, Sanekata T, Shibahara T, Inanaga S, Tanaka K (2007) Overexpression of monodehydroascorbate reductase in transgenic tobacco confers enhanced tolerance to ozone, salt and polyethylene glycol stresses. Planta 225:1255–1264
3. Eltayeb AE, Kawano N, Badawi GH, Kaminaka H, Sanekata T, Morishima I, Shibahara T, Inanaga S, Tanaka K (2006) Enhanced tolerance to ozone and drought stresses in transgenic tobacco overexpressing dehydroascorbate reductase in cytosol. Physiol Plant 127:57–65

Chapter 6
Distribution and Transport of Ascorbate

The ascorbate content in plants varies with different tissues and plant species. The ascorbate concentration in pepper is generally higher than that in tomato [1], and the ascorbate content in leaves is much higher than that in fruits [2]. Photosynthetic tissues as well as fruits and other storage organs usually contain relatively higher concentration of ascorbate [3]. Ascorbate content in rapidly-growing tissues is higher than that of aging tissues, as ascorbate is generally accumulated in the tissues with active growth such as the meristem [4]. This is consistent with the fact that content of ascorbate decreases with plant growth in most of parts of dill plants [5]. The ascorbate content in plants also varies with developmental stages [6]. The ripening fruits in tomato accumulate more ascorbate than immature fruits. The ascorbate content is reported to be affected by plant physiological status and environmental factors [7].

For the subcellular distribution, ascorbate is present in all the cell compartments of higher plant cells including cell wall [8, 9]. The average content of ascorbate in plant cells is around 2–25 mmol/L [10], and chloroplasts can accumulate high concentration (over 20 mmol/L) of ascorbate as much as 12–30 % of the total ascorbate content in leaves [8, 11]. Ascorbate is also present in the apoplast [12], cytoplasm and vacuole [13]. The cytoplasmic ascorbate pool size is estimated to be around 20 mmol/L [11]. There are also large amounts of ascorbate in mitochondria, as the final step catalysis enzyme for the ascorbate biosynthesis, GLDH, is located in the mitochondria [14]. The distribution of ascorbate in plant cells is presented in Fig. 6.1.

From the location of synthesis in the mitochondria, ascorbate must be transported to other cellular compartments where it accumulates to high concentrations. Ascorbate transportation in plants mainly includes two parts: the subcellular transportation between cellular compartments and extracellular transport to apoplast through the plasmalemma [15]. The presence of regulatory transport system of ascorbate allows rapid allocation among different cell sections responding to physiological, developmental and metabolic requirements of plants.

Y. Zhang, *Ascorbic Acid in Plants*, SpringerBriefs in Plant Science,
DOI: 10.1007/978-1-4614-4127-4_6, © The Author 2013

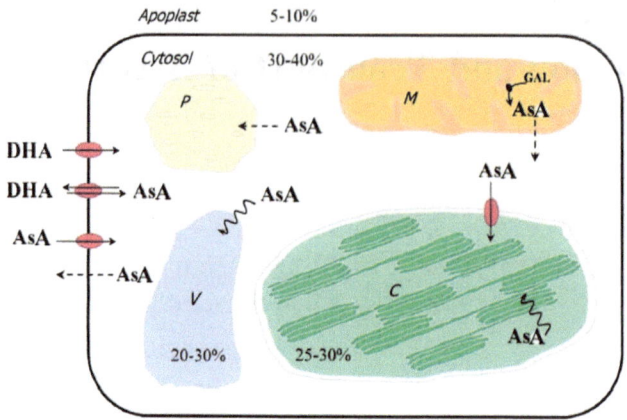

Fig. 6.1 Distribution and transport of ascorbate in plant cells (Cited from Horemans et al. 15, 16) *V* vacuole; *M* mitochondrion; *C* chloroplast; *P* peroxisome ──●▶ identified ascorbate (AsA) or dehydroascorbate (DHA) carrier systems; ∿∿ transport that is not carrier mediated; ----▶ translocation that has not been studied

It has been found that ascorbate transportation in plant cells can occur in the chloroplast, thylakoid, vacuole and plasmalemma, while transportation on mitochondrion and peroxisome is rarely reported. People believe that the ascorbate transfer from the mitochondrial inner membrane to the cytoplasm as the result of simple diffusion, as the final step of the ascorbate synthesis takes place on the mitochondrial inner membrane (Fig. 6.1). However, translocation of ascorbate through the plasmalemma and chloroplast membrane is mediated by specific carriers [16]. Ascorbate transportation from the cytoplasm to the chloroplast stroma is a carrier-mediated process, which may be driven by facilitated exchange diffusion. The ascorbate transportation on thylakoid membrane, however, is not a carrier-mediated process, but a pH and concentration gradient dependent simple diffusion [11].

There is evidence indicating that ascorbate/dehydroascorbate need to be transferred between the apoplast and the cytoplasm [17]. Apoplastic ascorbate amount accounts up to 10 % of the leaf total ascorbate content [12]. Due to the apoplastic lack of NADPH or glutathione, ascorbate can not be recycled by ascorbate–glutathione recycling in apoplast; therefore ascorbate has to be transferred from the cytoplasm to the apoplast through plasmalemma.

Up to present, there are at least three mechanisms driving the ascorbate/dehydroascorbate transport on plant plasmalemma: facilitated diffusion, proton electrochemical gradient transport and ascorbate/dehydroascorbate exchange transport. Initial observations indicate that carriers for both ascorbate and its oxidised form of dehydroascorbate are present in plant membranes, though transporters of ascorbate or dehydroascorbate have not been identified at the molecular level [18]. The ascorbate carrier on plant plasmalemma preferentially translocates the fully oxidized (dehydroascorbate) molecule [19]. Strong correlation existed between the level of ascorbate oxidation and the amount of transported molecules into the vesicles. The administering of ascorbate oxidants such as ascorbate oxidase results in a stimulated ascorbate uptake into the plasmalemma vesicles. Compared with the chloroplast membrane, ascorbate and dehydroascorbate carriers on the plasmalemma show a higher affinity.

The plasmalemma is both a bridge and a barrier between the cytoplasm and the outside world. Ascorbate transportation through plasmalemma in plants was firstly verified in soybeans [20], followed by evidences from isolated protoplasts of barley leaves [13], pea [11], bean [19] and tobacco BY-2 cell lines [21], which further confirmed the existence of ascorbate and dehydroascorbate transporters on the plasmalemma.

The apoplastic ascorbate–dehydroascorbate redox couple is linked to the cytoplasmic ascorbate–dehydroascorbate redox couple by specific transporters for either or both metabolites [15]. Parsons et al. [22] found that in response to exogenous hydrogen peroxide intracellular ascorbate in plant cell-suspension cultures is released and transported to apoplast, resulting in stimulated ascorbate and dehydroascorbate accumulation in the apoplast. This ROS stimulated transport is consistent with the role of ascorbate in signal perceiving and transduction as well as the antioxidant in apoplast. Apoplastic ascorbate is involved in a number of important physiological processes, including cell division, elongation and stress resistance, and therefore ascorbate transportation via plant plasmalemma is of huge physiological importance.

Although evidence about the mechanisms driving ascorbate or dehydroascorbate transport remains inconclusive, these carrier proteins potentially regulate the level and redox status of ascorbate in the apoplast. The redox coupling between compartments facilitated by these transport systems allows coordinated control of key physiological responses to environmental cues [15]. Regulation of ascorbate transport systems may be central in the regulation of different physiological processes including progression through the cell cycle, expansion of the cell wall and defence against abiotic and biotic threats [16].

Ascorbate transport is also demonstrated among different organs or tissues of plant to meet the demands of plant growth. Although all cells in plants have potentially the capacity to synthesize the ascorbate, the huge difference of ascorbate concentration in different tissues may indicate the synthesis capacity or the transport machinery among tissues. That the leaves usually contain the highest concentration of ascorbate among all plant organs may result from photosynthesis and its product accumulation. This will in turn help leaves with photoprotection. Evidence shows that the high concentration of ascorbate accumulated in source leaf are loaded into phloem and transported to sink tissues in plants [23]. Ascorbate is transported in the phloem, and glucose conjugates is found to occur in the phloem.

When [14C] ascorbate was applied to leaves of intact plants, autoradiographs and HPLC analysis demonstrated that ascorbate was accumulated into phloem and transported to root tips, shoots, and floral organs, but not to mature leaves.

Ascorbate was also directly detected in Arabidopsis sieve tube sap collected from aphid stylet, showing that the ascorbate was loaded in phloem. Feeding a single leaf with L-galactono-1,4-lactone, the immediate precursor of ascorbate, also resulted in significant increase in ascorbate concentration in the treated leaves, and moderate increase in untreated sink tissues of the same plant. The amount of ascorbate produced in treated leaves and accumulated in sink tissues was proportional to the amount of L-galactono-1,4-lactone applied. This result at least indicates that ascorbate or its precursor is transported among the plant organs. Studies of the ability of organs to produce ascorbate from L-galactono-1,4-lactone showed mature leaves have higher biosynthetic capacity and much lower ascorbate turnover rate than sink tissues. The results indicate ascorbate transporters from source leaves to sink tissues reside in the phloem, and that ascorbate translocation is likely required to meet ascorbate demands of rapidly growing non-photosynthetic tissues. Source leaf ascorbate biosynthesis is probably limited by substrate availability rather than biosynthetic capacity, and sink ascorbate levels may be limited to some extent by source production [23].

References

1. Ogunlesi M, Okiei W, Azeez L, Obakachi V, Osunsanmi M, Nkenchor G (2010) Vitamin C contents of tropical vegetables and foods determined by voltammetric and titrimetric methods and their relevance to the medicinal uses of the plants. Int J Electrochem Sci 5:105–115
2. Zhang CJ, Liu JX, Zhang YY, Cai XF, Gong PJ, Zhang JH, Wang TT, Li HX, Ye ZB (2011) Overexpression of *SlGMEs* leads to ascorbate accumulation with enhanced oxidative stress, cold, and salt tolerance in tomato. Plant Cell Rep 30:389–398
3. Loewus FA, Loewus MW (1987) Biosynthesis and metabolism of ascorbic-acid in plants. Crit Rev Plant Sci 5:101–119
4. Luwe MWF, Takahama U, Heber U (1993) Role of ascorbate in detoxifying ozone in the apoplast of spinach (*Spinacia oleracea* L) leaves. Plant Physiol 101:969–976
5. Lisiewska Z, Kmiecik W, Korus A (2006) Content of vitamin C, carotenoids, chlorophylls and polyphenols in green parts of dill (*Anethum graveolens* L.) depending on plant height. J Food Compos Anal 19:134–140
6. Birghila S, Dobrinas S, Matei N, Magearu V, Popescu V, Soceanu A (2004) Distribution of Cd, Zn and ascorbic acid in different stages of tomato (*Lycopersicum esculentum Solanaceae*) plant growing. Rev Chim-Bucharest 55:683–685
7. Smirnoff N (1996) The function and metabolism of ascorbic acid in plants. Ann Bot 78:661–669
8. Smirnoff N, Wheeler GL (2000) Ascorbic acid in plants: biosynthesis and function. Crit Rev Plant Sci 19:267–290
9. Zechmann B, Stumpe M, Mauch F (2011) Immunocytochemical determination of the subcellular distribution of ascorbate in plants. Planta 233:1–12
10. Davey MW, Gilot C, Persiau G, Ostergaard J, Han Y, Bauw GC, Van Montagu MC (1999) Ascorbate biosynthesis in Arabidopsis cell suspension culture. Plant Physiol 121:535–543
11. Foyer CH, Lelandais M (1996) A comparison of the relative rates of transport of ascorbate and glucose across the thylakoid, chloroplast and plasmalemma membranes of pea leaf mesophyll cells. J Plant Physiol 148:391–398

12. Vanacker H, Carver TLW, Foyer CH (1998) Pathogen-induced changes in the antioxidant status of the apoplast in barley leaves. Plant Physiol 117:1103–1114

13. Rautenkranz AAF, Li LJ, Machler F, Martinoia E, Oertli JJ (1994) Transport of ascorbic and dehydroascorbic acids across protoplast and vacuole membranes isolated from barley (*Hordeum vulgare* L cv Gerbel) leaves. Plant Physiol 106:187–193

14. Bartoli CG, Pastori GM, Foyer CH (2000) Ascorbate biosynthesis in mitochondria is linked to the electron transport chain between complexes III and IV. Plant Physiol 123:335–343

15. Horemans N, Foyer CH, Asard H (2000a) Transport and action of ascorbate at the plant plasma membrane. Trends Plant Sci 5:263–267

16. Horemans N, Foyer CH, Potters G, Asard H (2000b) Ascorbate function and associated transport systems in plants. Plant Physiol Biochem 38:531–540

17. Luwe M (1996) Antioxidants in the apoplast and symplast of beech (*Fagus sylvatica* L) leaves: seasonal variations and responses to changing ozone concentrations in air. Plant Cell Environ 19:321–328

18. Smirnoff N (2011) Vitamin C: the metabolism and functions of ascorbic acid in plants. Adv Bot Res 59:107–177

19. Horemans N, Asard H, Caubergs RJ (1997) The ascorbate carrier of higher plant plasma membranes preferentially translocates the fully oxidized (dehydroascorbate) molecule. Plant Physiol 114:1247–1253

20. Mozafar A, Oertli JJ (1993) Vitamin-C (ascorbic-acid): uptake and metabolism by soybean. J Plant Physiol 141:316–321

21. Horemans N, Potters G, Caubergs RJ, Asard H (1998) Transport of ascorbate into protoplasts of nicotiana tabacum bright yellow-2 cell line. Protoplasma 205:114–121

22. Parsons HT, Fry SC (2010) Reactive oxygen species-induced release of intracellular ascorbate in plant cell-suspension cultures and evidence for pulsing of net release rate. New Phytol 187:332–342

23. Franceschi VR, Tarlyn NM (2002) L-ascorbic acid is accumulated in source leaf phloem and transported to sink tissues in plants. Plant Physiol 130:649–656

Chapter 7
Enzymes Involved in Ascorbate Biosynthesis and Metabolism in Plants

Although several pathways for ascorbate biosynthesis have been successively proposed in plants, more evidence from biochemistry and genetics indicates that Smirnoff pathway is the major mechanism by which plants synthesize ascorbate. With the cloning of the gene encoding the last step enzyme in Smirnoff pathway in 2007, all the genes encoding for the key enzymes in Smirnoff pathway have been identified in Arabidopsis.

7.1 Phosphomannose Isomerase (PMI: EC 5.3.1.8)

Phosphomannose isomerase (PMI) catalyzes reversible conversion between fructose-6-phosphate and mannose-6-phosphate, which is first enzymatic step from phosphohexose to D-mannose. PMI was early investigated in microorganism. PMI from *E. coli* can confers mannose tolerance to plants, avoiding the potential biosafety concern related to conventional selectable marker genes, such as antibiotics or herbicide resistance genes. *PMI* was widely utilized as alternative selectable marker gene in various crop plants, including maize [1], lettuce [2], rapeseed [3], rice [4], tomato, potato [5], wheat [6], apple [7], Chinese cabbage [8], and fiberflax [9].

Research on phosphomannose isomerase in plants makes slow progress as compared to that in microorganism. The gene encoding phosphomannose isomerase in Arabidopsis was not cloned until 2008 [10]. Database mining of NCBI has generated two *PMI* genes from Arabidopsis. They encoded PMI enzyme with the conserved domain of YXDXNHKPE without any signal peptides. The PMI enzyme activity was inhibited by EDTA, Zn^{2+}, Cd^{2+} and ascorbate. *AtPMI1* and *AtPMI2* exhibited different expression profiles. *AtPMI1* was constitutively expressed in various tissues of Arabidopsis under normal growth conditions, and the gene expression can be induced by continuous light. Expression of *AtPMI2* could only be induced by long duration in dark. Correlation exists between gene expression of *AtPMI1* and PMI enzyme activity and ascorbate accumulation

Y. Zhang, *Ascorbic Acid in Plants*, SpringerBriefs in Plant Science, DOI: 10.1007/978-1-4614-4127-4_7, © The Author 2013

throughout different time points of the day [10]. Further investigation on the *AtPMI1* RNAi transgenic plants and the *AtPMI2* T-DNA knock-out line indicated that, decreased expression of *AtPMI1* led to reduced total ascorbate content in leaves, while *AtPMI12* expression decreasing did not show any effect on ascorbate accumulation in leaves. These indicate that the *AtPMI1*, but not *AtPMI2*, is involved in ascorbate biosynthesis in Arabidopsis. The authors argued *AtPMI2* is possibly involved in the utilization of mannose-derived carbohydrates as an energy source under sugar-starved conditions, such as prolonged darkness and/or senescence [10].

7.2 Phosphomannomutase (PMM: EC 5.4.2.8)

Phosphomannomutase catalyzes the interconversion of mannose-6-phosphate to mannose-1-phosphate in the D-Man/L-Gal pathway for the biosynthesis of ascorbate. The enzyme function has been widely investigated in fungi and mammals. Although the PMM enzyme activity has been detected in several plant species like red alga [11], maize [12] and spinach [13], the gene encoding PMM was not identified in plant species until recently. Qian et al. isolated the full length cDNAs of *PMM* gene from Arabidopsis, tobacco, soybean, tomato, rice and wheat. Amino acid comparison indicates plant-origin PMM share over 50 % identity with counterparts of human and fungi. The plant PMM consists of about 250 amino acids, and contains the conserved domain of DVDGT found in humans and yeast. In line with the similarity in primary structure, plant PMM complemented the sec53-6 temperature sensitive mutant of yeast. Arabidopsis PMM protein expressed in *E. coli* can catalyze the conversion from mannose-1-phosphate to mannose-6-phosphate and glucose-1-phosphate to glucose-6-phosphate, with the former reaction being more efficient than the later one [14].

PMM is constitutively expressed in Arabidopsis and tobacco. Virus-induce gene silencing of *PMM* in tobacco led to the significant decreasing of ascorbate content in tobacco leaves. And viral-vector-mediated ectopic expression of *PMM* in tobacco resulted in 20–50 % increasing of ascorbate content [14]. These research results indicate that PMM is involved in regulating the ascorbate biosynthesis, which was further supported by transgenic validation. Overexpressing of *PMM* gene in Arabidopsis increased the ascorbate by 25–33 % as compared to the wild-type control [14]. Thus PMM is proved to be involved and play a pivotal role in ascorbate biosynthesis and metabolism in tobacco and Arabidopsis.

The ascorbate content is found to be correlated with the *PMM* gene expression in the ripening fruits and leaves of acerola (*Malpighia glabra*) [15]. Badejo et al. overexpressed the *PMM* cDNA from acerola in tobacco, and the ascorbate content in transgenic tobacco increased up to 2 folds as compared to the untransformed control, with a corresponding correlation with the PMM transcript levels and activities [15]. Also they found that PMM enzyme activity correlated with ascorbate accumulation in acerola, tomato and Arabidopsis.

7.3 GDP-D-Mannose Pyrophosphorylase (GMP: EC 2.7.7.22)

GDP-D-mannose pyrophosphorylase (GMP) is localized in cytoplasm, catalyzing the synthesis of GDP-D-mannose from GTP and mannose-1-phosphate. GMP is an important enzyme for both eukaryote and prokaryote, and is regarded as the potential rate-limiting enzyme for the Smirnoff pathway of ascorbate biosynthesis [16].

GMP is a relatively upstream enzyme in the ascorbate synthesis pathway, catalyzing the formation of ascorbate synthesis intermediate GDP-D-mannose. *GMP* gene was firstly cloned by map based method from ascorbate defective mutant *vtc1* in Arabidopsis, which is also one of the earliest isolated genes in ascorbate biosynthesis pathway [17]. Other *GMP* genes have been successively identified in potato [18], tomato [19], and acerola [20].

Screening a collection of ascorbate-deficient mutants of Arabidopsis identified *vtc1* mutant, which contains approximately 30 % ascorbate content of wild-type. Biochemical, molecular, and genetic evidences demonstrated that the *vtc1* locus encodes a GDP-mannose pyrophosphorylase (mannose-1-P guanyltransferase). This enzyme acts to produce GDP-mannose, which is used for cell wall carbohydrate biosynthesis and protein glycosylation as well as for ascorbate biosynthesis [17].

In the *vtc1* mutant, DNA mutation at site +64 of *GMP* gene leads to the amino acid conversion from Proline to Serine at site 22. Although the mutation has not affected so much on the gene expression, the mutation leads to significantly reduced GMP activity, and thereby leaf ascorbate content is reduced to 30 % of that in wild-type. Overexpression of wild-type *GMP* gene in the *vtc1* could restore the ascorbate accumulation in leaves [17, 21]. Methyl jasmonate can be utilized as an inducer to up-regulate *GMP* genes to improve the ascorbate content in Arabidopsis.

Badejo et al. found out that in acerola, West Indian cherry (*Malpighia glabra*), with high ascorbate content, the *MgGMP* gene expression correlated with ascorbate accumulation. Overexpressing the *MgGMP* gene along with its upstream promoter in tobacco improved the ascorbate content up to two folds of the wild-type control [22]. *GMP* gene is highly expressed in tomato leaves, consistent with the higher ascorbate content in tomato leaves. It is also found that high correlation occurred between *GMP* gene expression and ascorbate content in Arabidopsis [22]. It is also reported that antisense inhibition of *GMP* gene in transgenic potato resulted in significantly reduced ascorbate content in leaves and tubers [18].

Alteration of the expression of *SlGMP* significantly changed the total ascorbate content in tomato. The ascorbate level in *SlGMP* over-expressing transgenic line increased 22 % in leaves and 60 % in fruits, and the *SlGMP* RNAi transgenic line showed 26 and 29 % decreasing ascorbate level in leaves and fruits, respectively, as compared to the wild-type. These results showed that *SlGMP* plays an important role in regulating ascorbate biosynthesis in tomato (unpublished data). The expression profile in break and red ripening fruits of *SlGMP* over-expressing plants were analyzed by cDNA microarray, and the results showed that genes involved in cell component, lipid metabolism, ethylene biosynthesis and stress response were up-regulated in both break and red ripening fruits (unpublished data). The *GMP* overexpression in tomato did not affect the expression of other genes in the ascorbate biosynthesis, suggesting that the ascorbate content improvement was mostly caused by *GMP* gene (unpublished data).

GMP catalyses the biosynthesis of GDP-D-mannose, an important step in the formation of all guanosine-containing sugar nucleotides in plants which are precursors for cell wall biosynthesis as well as synthesis of ascorbate. Therefore, GDP-mannose participates in numerous cellular processes, such as non-structural sugar synthesis in cell wall as well as ascorbate biosynthesis in plants [17]. The mannose of the hemicellulose polymeride such as glucomannan and galactomannan comes from the activated GDP-mannose. The GDP-L-trehalose in the cell wall and glycoprotein are also derived from GDP-mannose under the catalyzing of GDP-D-mannose-4, 6-dehydratase. The trace component of L-galactose in plant cell wall is also synthesized from GDP-mannose. At the same time, GDP-mannose plays an important role in the process of protein glycosylation. The major component of the O-linked glycoprotein and N-linked glycoprotein, D-mannose, originates from GDP-mannose [17].

In another *GMP* mutant in Arabidopsis, *cyt1*, the lack of GDP-mannose leads to series of secondary reactions, including incomplete cell wall due to altered cell wall components (lowered content of mannose, trehalose and cellulose in the mutant), lack of N-linked glycosylation and decreasing accumulation of ascorbate [23]. The transgenic potato plants with antisense *GMP* gene exhibited lowered concentration of ascorbate, and premature senescence with numerous brown spots on the stems and leaves, though the antisense transgenic plants were indistinguishable from the wild-type under tissue culture conditions. The mannose concentration in the cell wall of transgenic plant leaves decreased to 30–50 % of untransformed control, while the protein glycosylation pattern remained unchanged [18]. Compared to the wild-type control, *vtc1* mutant also showed altered growth pattern such as less plant height, delayed florescence, and presenility [24].

Recently several other reports showed that the *vtc1* mutant was extremely sensitive to ammonium (NH_4^+), which is supposed to be caused by the lack of N-glycosylation [25–27]. It has been shown that the product of the GMP catalyzing, GDP-D-mannose, is involved in cell wall synthesis and protein glycosylation as well as the ascorbate synthesis. Synthesis of moderate amount of GDP-D-mannose under the catalyzing of GMP enzyme is indispensable in various plant biological processes, indicating the vast importance of GMP in plant kingdom.

Thus the present evidence indicates GMP enzyme-catalyzed reaction step is vital to ascorbate synthesis and plant development. Although GMP catalyzing step is indispensable and quite effective for the regulation of ascorbate synthesis, whether it's the most critical regulatory step remains inconclusive.

7.4   GDP-D-Mannose-3,5-Epimerase (GME: EC 5.1.3.18)

GDP-D-mannose-3,5-epimerase (GME) is another one of the most important enzymes in the process of ascorbate biosynthesis, catalyzing the conversion from GDP-D-mannose to GDP-L-galactose. The GME enzyme was firstly identified from *Chlorella pyrenoidosa*, pea and Arabidopsis [16]. In 2001, Wolucka et al. isolated the highly purified GME enzyme from cell suspension of Arabidopsis through combination of conventional biochemistry and mass spectrum analysis [28]. And the gene encoding GME enzyme was then cloned by trypsin digestion, polypeptide sequencing, and comparison analysis. GME expressed in *E.coli* cells exhibited the enzyme activity. GME enzyme exists in form of homodimers with molecular weight of 84 kD. At the N terminal of the GME protein there is typical binding site of NAD(P), namely the glycine-rich GXGXXG domain. GME enzyme also possesses the property of short-chain dehydrogenases/reductases, with the conserved serine/threonine residues and YXXXK motifs, which jointly comprise the highly conserved catalyzing site of GME enzyme. NAD(P)$^+$ can improve the activity of GME, while NAD(P)H inhibits GME activity [29].

Co-purification of GME and Hsp70 was observed both in *E.coli* and Arabidopsis cell suspension during the process of GME enzyme purification, which may suggest that Hsp70 molecular chaperones could possibly help in protein folding and regulation of GME enzyme [29]. Other investigators argue that interaction between heat shock proteins and GME enzyme could be the regulation mechanism of plant to overcome the environmental stresses [30].

The GME catalyzes at least two distinct epimerization reactions and releases, besides the well known GDP-L-galactose, a novel intermediate of GDP- L-gulose. In addition to catalyzing the synthesis of GDP-L-galactose from GDP-D-mannose, GME also helps to synthesize another product of GDP-L-gulose, which is considered as intermediate in the gulose pathway of ascorbate biosynthesis in plants.

GME genes have been isolated and characterized in several high plant species, such Arabidopsis [28, 29], rice [31] and tomato [32]. GME exists in form of single copy in most of plant species, while two *GME* family members were found in rice and tomato, respectively [31]. Zou et al. localized the tomato *GME2* gene onto the 9-J locus using the IL population by RFLP approach [19]. QTL analysis of the genes for the ascorbate accumulation in red ripening fruit using three different tomato genetic populations, taken together results of Zou et al. showed that 9-J domain with *GME2* was the common QTL locus among the three populations, suggesting GME is one of the most important enzymes for ascorbate biosyntheses in tomato [33].

Investigation on the *GME* gene through overexpression or gene inhibition was not widely reported so far. Until 2009, Gilbert et al. investigated the *GME* function in tomato through the RNA interference approach [34]. When the two *GME* genes were suppressed simultaneously in tomato, transgenic plants showed deceasing ascorbate concentration, ROS accumulation, chlorosis and bleaching of leaves, and growth retardation. Further evidence showed significant changes in cell wall components in the *GME* suppressed plants. In all the stems and 20 day-post-florescence fruits detected, the proportion of mannose in cell wall components increased while the galactose proportion decreased as compared to untransformed control. The mechanical property of transgenic plants was also altered as the cell wall components changed, with the leaves and stem being more fragile compared to control plants. Based on those observations, the authors proposed that besides its important role in ascorbate biosynthesis, GME is involved in formation of non-cellulosic component of cell walls [34].

GME exists as a single copy gene in most crop species, while two copies of *GME* were found in rice and tomato genome. Belonging to the same gene family, *SlGME1* and *SlGME2* show many similar characteristics, e.g. similar gene structure, the same base number of the exons, the same number of amino acids encoded, and the same localization in the cytoplasm. The two members of *GME* gene family in tomato share a 92 % similarity with each other. However, *SlGME2* contains two larger introns as compared to *SlGME1*. A similar case is found in rice, where *OsGME2* have three larger introns compared with *OsGME1*. The *SlGME1* and *SlGME2* are both constitutively expressed in the various tomato tissues with different expression levels in fruits. *SlGME1* is expressed at a relatively consistent abundance in different fruit development stages, while the expression of *SlGME2* decreases as the fruit development and ripening. In the leaves of transgenic tomato overexpressing *SlGME2*, *SlGME1* expression level was also enhanced as compared to wild-type control, indicating the possible mutual regulation between *SlGME1* and *SlGME2* (unpublished data).

Overexpression of either *SlGME1* or *SlGME2* leads to improvement of the ascorbate accumulation in leaves and fruits in tomato compared with untransformed control. *GME* overexpressing tomato plants also showed less chlorophyll loss and membrane-lipid peroxidation under methyl viologen (paraquat) stress, higher survival rate under cold stress, and significantly higher germination rate, fresh weight, and root length under salt stress compared to the wild-type control, suggesting that *GME* gene plays an important role in stress tolerance as well as in ascorbate synthesis [32]. Gene expression profile in break fruits of *SlGME1* overexpressing plants were analyzed by cDNA microarray and the results indicated that several genes involved in stress response, lipid metabolism and flavonoids biosynthesis were up-regulated. A few genes with altered expression profile in break fruits of either *SlGME1* or *SlGME2* overexpressing plants showed the same change trend (unpublished data).

Among the *SlGME2* sense transgenic tomato plants, several co-suppression lines with *SlGMEs* suppression and ascrobate decreasing were identified. The *SlGME2* co-suppression transgenic plants with the *SlGMEs* inhibition showed significant decreasing of ascorbate concentration in leaves and fruits. The total ascorbate content in leaves and break fruits of *SlGME2* co-suppression lines was decreased to 12 and 28 % of that in wild-type plants, respectively, suggesting that *SlGMEs* effectively modulate ascorbate biosynthesis in the tomato. It is interesting that *SlGME2* co-suppression transgenic tomato plants showed apical shoots retardation, which is possibly attributed to the extremely reduced ascorbate concentration in leaves (1.9–15.0 % of wild-type control). The gene expression analysis showed that the expression level of *SlGME1* but not *SlGME2* was drastically inhibited in the *SlGME2* co-suppression line as compared to the wild-type control, resulting in substantial deprivation of ascorbate synthesis capacity. It can be inferred that *SlGME1* may play a greater role than *SlGME2* in the regulation of leaf ascorbate biosynthesis. However, this speculation needs to be verified by gene knock-out of the two members of the *GME* genes respectively in tomato.

In addition, both overexpression and suppression of *SlGMEs* altered the transcriptional abundance of several genes related to ascorbate biosynthesis. These results indicated that *SlGMEs* are effectively involved in the regulation of tomato ascorbate biosynthesis [32].

7.5 GDP-L-Galactose Phosphorylase (GGP: EC 2.7.7.69)

GDP-L-galactose phosphorylase catalyzes the synthesis of L-galactose-1-phosphate from GDP-L-galactose. The gene encoding GDP-L-galactose phosphorylase is the last isolated gene involved in Smirnoff pathway for ascorbate biosynthesis. Three defective mutants related to the gene, *vtc2-1*, *vtc2-2* and *vtc2-3,* were screened out in 2000 [35]. *VTC2* gene was cloned using map-based method in 2002, however the gene sequence comparison from database only generated the unknown proteins due to the lack of gene annotation in the database. Thus how the cloned *VTC2* gene was involved in ascorbate biosynthesis remained unknown until it was functionally characterized in 2007. Sequence comparison analysis shows *VTC2* gene belongs to a histidine triad superfamily. The prokaryotic expressed VTC2 protein from Arabidopsis and kiwifruit in *E. coli* could catalyze GDP-L-galactose into L-galactose-1-phosphate. The transiently expressed kiwifruit VTC2 in tobacco leaves could improve the ascorbate in tobacco leaves over 31 folds, and the GDP-L-galactose-D-mannose-1-phosphate guanyltransferase activity was increased to 50 folds. The VTC2 was proved to be the gene responsible for the reaction step from GDP-L-galactose to L-galactose-1-phosphate in the ascorbate biosynthesis [36]. At the same time, the evidence from sequence similarity and enzyme activity indicated that the enzyme encoded by *VTC2* is a member of the GalT/Apa1 branch of the histidine triad protein superfamily that catalyzes the conversion of GDP-L-galactose to L-galactose 1-phosphate, which adds to elucidation of VTC2 function [37]. Following the characterizing recombinant VTC2 from Arabidopsis as a specific GDP-L-galactose phosphorylase, all enzymes catalyzing each of the ten steps of the Smirnoff-Wheeler pathway from glucose to ascorbate have been identified.

Another homologues gene with *VTC2*, *At5g55120,* was identified in Arabidopsis [38]. Amino acid sequence analysis indicated that *VTC2* and *At5g55120* both encode hydrophilic proteins without transmembrane domains or organelle localized sequences. That means the encoded proteins are likely to be located in cytoplasm. T-DNA insertion mutants of *At5g55120* gene, *vtc5-1* and *vtc5-2,* showed decreasing in ascorbate accumulation. Seedlings of double mutants obtained by crossing *vtc2-1* with *vtc5-1* or *vtc5-2* exhibit growth retardation after germination, which could be recovered by supplemented ascorbate or L-galactose, suggesting the two *GGP* genes are both functionally important for the ascorbate biosynthesis in Arabidopsis. *GGP* gene was constitutively expressed throughout the plant growth and development, with the significant higher expression abundance in green tissues than that in roots [39]. The expression of two *GGP* genes in Arabidopsis are regulated by light, and the ascorbate accumulation increases as illumination intensity strengthens, indicating the *GPP* participates in the light-regulated process of the ascorbate biosynthesis [38].

The ascorbate content in leaves and different developmental stages of fruits of kiwifruits highly correlated with *GGP* gene expression. The gene expression level of *GME* and *GGP* in *A. eriantha*, the kiwifruit species with the highest ascorbate concentration, was higher than that in other two kiwifruit species. Overexpressing kiwifruit *GGP* gene in Arabidopsis resulted in 4-fold increase in ascorbate accumulation, while 7-fold increase in ascorbate was observed when *GME* and *GGP* are transiently co-expressed in Arabidopsis [40]. These results indicate *GGP* will likely be the critical step regulating the ascorbate biosynthesis in plants.

The mechanism of how GGP enzyme interacts with its substrate, GDP-L-galactose, still remains controversial. One hypothesis is that GGP enzyme transfers the GMP (guanosine monophosphate) from GDP-L-galactose to a kind of hexose-1-phosphate, which is likely to be D-mannose-1-phosphate [36]. Another totally different opinion argues that GGP enzyme may interact with GDP-L-galactose to produce L-galactose-1-phosphate by guanylylating its conserved active His residue [41].

Recent evidence indicates that the GGP might act as a dual-functional protein. Subcellular localization of Arabidopsis GGP showed unexpectedly that the fusion protein of GGP::GUS existed not only in cytoplasm, but also in the nucleolus, suggesting that GPP might function as a regulatory factor as well as catalyzing enzyme [39].

7.6 L-Galactose-1-Phosphate Phosphatase (GPP: EC 3.1.3.25)

L-galactose-1-phosphate phosphatase (GPP) catalyzes the transformation reaction from L-galactose-1-phosphate to L-galactose. The enzyme of L-galactose-1-phosphate phosphatase was firstly purified from young berry of kiwifruit in 2004, with the molecular weight of 65 kD [42, 43]. The GPP can specifically hydrolyze the L-galactose-1-phosphate. GPP enzyme shows the Mg^{2+} dependent activity, while surplus Mg^{2+} concentration over 2 mmol/L will inhibit its activity [43]. Due to the extreme instability of highly purified GPP enzyme from kiwifruit, the GPP was purified from Arabidopsis seedling instead. Trypsin digestion and LC–MS analysis helped to identify an enzyme accession in GenBank database with the annotation of myoinositol 1-phosphate phosphatase (GenBank accession no.: At3g02870). Ectopic expression of the homologous gene of At3g02870 from kiwifruit in *E. coli* cells indicated that the prokaryotic expressed enzyme could simultaneously hydrolyze L-galactose-1-phosphate and myoinositol 1-phosphate, and the catalyzing rate with L-galactose-1-phosphate was 14-fold higher than that with myoinositol 1-phosphate [42].

The *VTC4* gene was map-based cloned in 2006, and proved to be the exactly At3g02870 identified by Laing et al. [44]. The mutation of Pro → Leu at 92 site in the active domain of the VTC4 enzyme results in decrease in GPP enzyme activity in *vtc4-1* mutant, and the ascorbate content in *vtc4-1* was reduced to 50 % of that in wild-type plants. In two Arabidopsis T-DNA insertion lines of At3g02870, the ascorbate content as well as the L-galactose-1-phosphate phosphatase activity was significantly lower than that of wild-type. The similar decreasing in ascorbate content and VTC4 enzyme activity was also observed in *vtc4-1* mutant. Genetic complementary test showed that the site disrupted in T-DNA insertion mutant share the same locus with *vtc4-1* mutant [44]. The *vtc4-1* mutation leads to the alteration in L-galactose metabolism and the accumulated L-galactose-1-phosphate makes galactose more available for polysaccharide synthesis. Expression profiles of genes involved in ascorbate biosynthesis and metabolism among developmental stages in tomato fruit indicate that *GPP* may play a pivotal role in ascorbate biosynthesis and metabolism in tomato [45]. In the T-DNA insertion lines of *GPP* gene, the still existing ascorbate in plants indicates that other enzymes or biological pathways may take part in ascorbate biosynthesis [44].

Recent study indicates that VTC4 is a dual-function enzyme, catalyzing the biosynthesis of both myoinositol and ascorbate. Enzyme kinetics analysis of recombinantly expressed VTC4 protein of Arabidopsis gave new insight into the enzyme function. The recombinantly expressed enzyme of VTC4 could efficiently hydrolyze D-myoinositol 3-phosphate, the precursor of myoinositol synthesis, as well as the L-galactose-1-phosphate. Gas chromatograph analysis of *VTC4* T-DNA knockout line in Arabidopsis revealed that content of both ascorbate and myoinositol in the T-DNA mutant decreased to only 20–30 % of wild-type control [46]. Thus *GPP* may serve as a connection point for myoinositol pathway and Smirnoff pathway of ascorbate biosynthesis.

So far the transgenic research on *GPP* gene was not reported. The *GPP* gene was cloned in tomato (not published data) and overexpression and RNAi suppression work is now underway to further investigate the function of GPP enzyme.

7.7   L-Galactose Dehydrogenase (GalDH: EC 1.1.1.117)

L-galactose dehydrogenase (GalDH) catalyzes the oxidation of L-galactose within its C1 site and the formation of L-galactono-1,4-lactone, a irreversible conversion in which L-galactose is specifically involved for ascorbate biosynthesis. GalDH enzyme is considered to be localized in cytoplasm without any obvious localization or transportation signals. GalDH functions as dimmers with its cofactor of NAD^+.

Thus far the *GalDH* gene is isolated form kiwifruit, apple [42], Arabidopsis [47], and spinach [48]. GalDH exists in single copy in the genome of Arabidopsis and spinach.

The L-galactose dehydrogenase enzyme in high plants was firstly isolated from hypocotyls of pea. The Arabidopsis counterpart was also identified based on similarity search with the N-terminal amino acid sequence of the pea GalDH [47]. The Arabidopsis GalDH expressed in *E. coli* possessed relatively high activity of L-galactose oxidization, and relatively low activity for oxidation of L-trehalose and L-sorbose. Overexpressing the Arabidopsis *GalDH* gene in tobacco resulted in up to 3.5-fold increase in GalDH enzyme activity, while the ascorbate content in the transgenic tobacco leaves was not significantly changed [47]. Antisense inhibition *GalDH* did not affect the ascorbate content in Arabidopsis leaves under weak light, but resulted in slight decrease in ascorbate under strong light. However, the slight decreasing of ascorbate was disproportionable as compared to the sharp decreasing of GalDH enzyme activity [47]. The relative stable accumulation of ascorbate in transgenic plants with *GalDH* possibly results from feedback regulation upon GalDH enzyme. When the plants are confronted with disadvantageous condition (decreasing ascorbate as antioxidant), feedback regulation mechanism is supposed to be activated to compensate for ascorbate biosynthesis.

Similar feedback regulation hypothesis could be applied on the *GMP* RNAi transgenic plants, in which *GMP* suppression led to up-regulated expression of the major genes involved in ascorbate biosynthesis in D-Man/L-Gal pathway (unpublished data). Enzyme activity of spinach GalDH is inhibited by ascorbate. Ascorbate of 1 mmol/L concentration results in 41 % decreasing of GalDH enzyme activity [48], indicating the feedback regulation mechanism in the ascorbate biosynthesis.

The catalyzing efficiency of GalDH is fairly high in term of L-galactose transforming. Exogenous L-galactose could be converted efficiently to ascorbate, resulting relative low content of L-galactose in plants [16, 49]. Thus GalDH catalyzing reaction may not be the rate-limiting step for the ascorbate biosynthesis, though its enzyme activity is feedback regulated by ascorbate.

7.8 L-Galactono-1,4-Lactone Dehydrogenase (GLDH: EC 1.3.2.3)

L-galactono-1,4-lactone dehydrogenase (GLDH), the last step enzyme in the ascorbate biosynthesis pathway, catalyzes the oxidization of the L-galactono-1, 4-lactone at its C2/C3 sites and the final formation of ascorbate. The GLDH enzyme, localized on mitochondrial inner membrane, utilizes the cytochrome c as its electron acceptor.

Thus far, the GLDH enzyme has been purified and characterized in several plant species, including sweet potato [50, 51], cauliflower [13, 52], spinach [53], tobacco [54], strawberry [55], melon [56], tomato [19, 57] and Arabidopsis (58).

GLDH enzyme was firstly purified from sweet potato [50], which is also one of the earliest characterized enzyme responsible for ascorbate biosynthesis, and the gene encoding GLDH was for the first time isolated in cauliflower [13]. *GLDH* is found to be single copy in the genomes of sweet potato, cauliflower, tobacco, melon, and Arabidopsis.

The amino acid sequence of GLDH is high conserved, resulting in similar enzyme properties among plant species. GLDH exists in form of monomer, with the molecular weight of 56 kD-57 kD. GLDH functions highly specifically on L-galactono-1,4-lactone, and its enzyme activity is inhibited by high concentration of L-galactono-1,4-lactone. Lycorine was also reported to inhibit the GLDH enzyme activity [59]. As for GLDH catalyzing mechanism, recent evidence indicates that the replacement of histidine with leucine (Leu56) may play a crucial role in both FAD-binding and catalysis [60].

The expression study of melon *GLDH* showed that *GLDH* was constitutively expressed in various melon tissues, without response to the environmental factors like hormone or stresses except light [56]. The *GLDH* gene in tobacco weakened with the increasing of exogenous ascorbate, and was up-regulated by light [61].

Antisense inhibition of *GLDH* gene expression in tobacco cell suspension BY2 resulted in a decrease of ascorbate content (70 % of wild-type control), and slower cell growth rate [62]. On the other hand, overexpression of *GLDH* in tobacco cell suspension BY2 led to enhancement in cell mitosis, senescence resistance, and anti-oxidization as well as the improved ascorbate content [63]. More recently, overexpressing the *GLDH* from sweet potato in tobacco, however, did not show any improvement in ascorbate accumulation in transgenic tobacco leaves [64]. The different effect of *GLDH* overexpressing in tobacco cell suspensions and plant remains to be elucidated.

The RNA interference mediated suppression of *GLDH* resulted in decrease in GLDH enzyme activity in tomato, but the ascorbate content in both leaves and fruit of transgenic tomato remained unchanged, showing the complexity of ascorbate regulation [65], which was further supported by the transgenic tobacco expressing *GLDH* [64]. Transgenic tobacco overexpressing potato *GLDH* gene exhibits 6- to 10-fold higher GLDH activities in the roots than the non-transformed plants. Despite the elevated GLDH activity, the ascorbate content in the leaves did not change significantly. Upon feeding substrate of L-galactono-1,4-lactone, ascorbate content in leaves of *GLDH* overexpressing plant as well as wild-type plants was improved with the same folds, which indicate that L-galactono-1,4-lactone may not be the rate-limiting step for ascorbate biosynthesis. Thus the critical step for ascorbate biosynthesis may reside upstream of GLDH [64].

This is reminiscent of the finding that overexpressing gene encoding myoinositol oxygenase (*MIOX*) in tomato did not improve the ascorbate content under normal growth condition, but promoted the ascorbate accumulation under inositol feeding (unpublished data). It is proposed that MIOX catalysis may not be a crucial step for ascorbate synthesis possible due to the low concentration of endogenous substrate.

The GLDH suppressed tomato also showed dwarf plant, smaller leaves and fruits, which was possible indicator that GLDH might participate in cell growth and development [65]. The relatively high expression of *GLDH* in expanding fruits of strawberry and melon, provided further evidence that GLDH may take part in regulating the cell growth and development [55, 56]. GLDH is also indispensable for proper assembly of respiratory chain complex [66].

7.9 Ascorbate Oxidase (AO: EC 1.10.3.3)

Ascorbate oxidase (AO) is a plant blue-copper protein catalyzing dioxygen reduction to water using ascorbate as the electron donor, by which ascorbate is oxidized into monodehydroascorbate [67]. AO is a glycoprotein with the majority of activity on the cell wall, indicating that AO is a cell wall enzyme [68, 69]. AO exists in all higher plants with a high level of AO activity in the rapid-growing plant tissues, such as fast-expanding fruits of the gourd family and germinating seeds. The AO mRNA and protein levels, as well as AO enzyme activity decrease when the rapid expansion of fruits comes to end [68].

Thus far, the gene encoding AO enemy was cloned from cucumber [70, 71], pumpkin [68], melon [72, 73], tobacco [74], tomato [18, 56] and several other plant species. A gene family of *Bp10* with high sequence similarity to cucumber and pumpkin ascorbate oxidase was cloned from rapeseed. However, the AO active centres are not conserved in the Bp10 products, suggesting an evolutionary relationship but a different enzymatic activity for these proteins [75].

AO gene expression is subjected to complicated transcriptional and translational regulation, and is affected by light and auxin. *AO* mRNA abundance in tobacco increases under light and auxin treatment, while salicylic acid reduces the *AO* transcription abundance [76].

It is widely considered that the physiological function of the AO lies in its importance in cell elongation [69, 77]. Overexpression of the pumpkin *AO* gene in tobacco BY-2 cell protoplast resulted in faster cell elongation than untransformed control. AO regulates cell expansion possibly through perceiving and transmitting environmental clues through plasmalemma [74]. It is also argued that the dehydroascorbate promotes the cell wall relaxation, resulting in cell elongation.

AO function in ascorbate metabolism was characterized by transgenic approaches. The change in *AO* activity did not affect the entire ascorbate pool in tobacco leaf, but the apoplastic ascorbate pool was significantly altered. Both sense and antisense expression of *AO* gene resulted in improved apoplastic ascorbate content (ascorbate + dehydroascorbate) as compared to untransformed control. The AO was proved to affect the redox state of apoplastic ascorbate. Sense *AO* expression led to the 40-fold increase in the apoplastic AO activity and 3.5-fold increase in dehydroascorbate content, and promoted plant growth and stem elongation. The antisense *AO* expression resulted in 2.5-fold decrease in AO enzyme activity and 42 % increase in the reduced ascorbic acid [78]. Overexpression of cucumber ascorbate oxidase in tobacco resulted in 380-fold increase in the enzyme activity. Although the total ascorbate content in apoplast and symplast of *AO* transgenic tobacco was not altered, the redox state of ascorbate was changed, accompanied by decreasing content of reduced ascorbic acid and increased sensitivity to ozone [73].

The role of AO on plant stress response was also investigated [79]. The AO enzyme activities of sense and antisense *AO* transgenic plants were 16-fold and 0.2-fold of wild-type plants, respectively. The transgenic plants did not show any phenotype changes under normal growth conditions. However, the germination rate, photosynthetic rate and seed yield of antisense transgenic plants were higher than that of sense transgenic and wild-type plants under high salt conditions. The reduced state of ascorbic acid in the sense transgenic plants were very low even under normal growth conditions, with the highest hydrogen peroxide concentration in both symplastic and apoplastic space of the sense transgenic plants. Arabidopsis T-DNA insertion mutant showed extremely low AO activity, with the similar phenotype of antisense *AO* transgenic tobacco plants [79], indicating that inhibition of apoplastic *AO* gene expression can increase plant salt tolerance.

AO gene expression was down-regulated to effectively enhance ascorbate accumulation in plants. The *AO* gene was RNA interference-mediated silenced in tomato, leading to decreased AO enzyme activity and significantly improved ascorbate content in tomato fruit. The decreasing AO enzyme and improved ascorbate were correlated with *AO* gene suppression. The AO suppression and ascorbate improvement was also accompanied by higher photosynthesis capacity as compared to untransformed control under drought stress [57]. This suggests a promising approach to improve ascorbate content in vegetable crops by regulating genes related to ascorbate metabolism.

It is interesting that the plant AO activity is relatively stable and even retains activity in the digestive system of the herbivore. And the AO activity in the digestive system will cause oxidative loss of ascorbate to herbivore. The oxidation of ascorbate also produces active oxygen species such as the highly reactive hydroxyl radical. So ascorbate oxidase may act as a plant defense protein against herbivory insects [80].

7.10 Ascorbate Peroxidase (APX: EC 1.11.1.11)

Ascorbate peroxidase (APX) utilizes the ascorbate as an electron donor to reduce hydrogen peroxide into water, producing monodehydroascorbate. It has been more than 30 years since APX was discovered. A peroxidase using ascorbate as electron donor was firstly discovered in 1976. In 1980, a peroxidase using photoreductant as an electron donor was reported in the intact spinach chloroplast, while the enzyme activity was very low in broken chloroplast. This peroxidase was later formally identified as APX, existing in the chloroplast stroma. Thereafter, huge efforts have been made on its enzymatic properties, distribution, location, function mechanisms, and molecular characteristics.

Ascorbate peroxidase is a hydrogen peroxide-scavenging enzyme that is specific to plants and algae and is indispensable to protect chloroplasts and other cell compartments from damage by hydrogen peroxide and hydroxyl radicals produced from it [81]. APX is one of the important components of plant ascorbate–glutathione cycle. As an important part of the antioxidative system, APX maintains the balance and uninterrupted functioning of the plant cell. The main role of APXs is to control the hydrogen peroxide concentration in cells. Due to the lack of catalase (CAT) and glutathione peroxidase (GPX) in the chloroplast, and the far less Km value for hydrogen peroxide of chloroplastic APX compared to CAT, APX is the key player in removing of hydrogen peroxide in chloroplasts.

APX activity has been observed in leaves of vast majority of plant species such as spinach, pea, and tomato, and in legume nodules, castor seeds, potato tubers and algae. Plant APX is composed of several isozymes. APX is distributed in at least four different cell compartments, after which APX is named: the cytosolic APX (cAPX), the stromal APX (sAPX) and thylakoid membrane-bound APX (tAPX) in the chloroplast, microbody (including glyoxysome and peroxisome) membrane-bound APX (mAPX), and mitochondrial membrane-bound APX (mitAPX) [81].

Chloroplastic and cytosolic isozymes of ascorbate peroxidase show differences in enzymatic properties [82]. Depletion of the electron donor ascorbate causes rapid inactivation of chloroplastic APX of higher plants, while cytosolic APX is stable under such conditions. The different isozymes of APX vary with the sensitivity to the depletion of ascorbate, which can be utilized to separately assay the APX isozymes in plants [83].

APXs belong to class I of the superfamily of bacterial, fungal and plant peroxidases. The active site in APX is highly conserved among APX family [84]. The evolution analysis based on the amino acid sequence of active catalytic domain of APX isozymes shows that APX isozymes of higher plants and algae can be divided into four groups, namely cAPX I and cAPX II, chlAPX, and mAPX. cAPXI, chlAPX and mAPX share common features conserved among plant species, while the cAPXII group may have evolved from cAPX in species-specific manner. In higher plants, APX isozymes show higher homology within the group (up to 70–90 %) as compared to the homology of 50–70 % between the different groups. Residues of Arg-38, His-42, Glu-65, Asn-71, His-163 and Asp-208 in all of the APX isozymes are highly conserved, which are indispensable for the binding of heme ligand. Trp-179 residue is conservative in most APX isozymes, participating in hydrogen bonding network together with His-163 and Asp-208 residues.

The different isozymes of APX are supposedly formed via the alternative splicing of APX pre-mRNA in higher plants. It is reported that four kinds of mature functional chloroplast APX isozymes are formed through alternative mRNA splicing: tAPX I, sAPX I, sAPX II and sAPX III. The expression ratio of tAPX to sAPX varies with different plant species. Alternative splicing of chloroplast APX pre-mRNA is a major regulatory mechanism to modulate the expression ratio of tAPX to sAPX in different tissues [81]. APX research on pumpkin cotyledon shows that light to some extent regulates the alternative splicing of the chloroplastic APX isozymes. The alternative splicing events in the

3'-terminal region of chloroplastic APX pre-mRNA were also regulated in a tissue-specific manner [85].

At present, the genes encoding APX isozymes have been successfully cloned from various plant species, such as pea, strawberry, tomato, potato, radish, Arabidopsis, rice, and alga [19, 86–93]. The pea *cAPX* gene (APX I) contains nine introns with the 5'-untranslated region in the first intron. Similar gene structure is also observed in other plant species. Several kinds of oxidization stresses can stimulate *cAPX* expression, indicating possible cis-regulatory elements in its promoter. High temperature induces *cAPX* gene expression in rice, and high temperature pre-treatment helps to confer chilling resistance in rice. Functional G/C-rich elements were found in the promoter of the *APX*, which is required for the ethylene induction. In addition, a 5' leader intron in tomato *APX* gene is essential for its constitutive expression in leaves [89].

Hydrogen peroxide can regulate the expression of *APX* gene. The hydrogen peroxide levels in plant cells rise in response to different stress conditions. Hydrogen peroxide is considered to be linked to plant MAP kinase pathways, indicating the role of hydrogen peroxide as signal in stress responses [94]. Exogenous hydrogen peroxide treatment resulted in altered transcriptional level of *cAPX* gene in soybean cell cultures. Treatment with the APX inhibitor, hydroxyurea, or CAT inhibitor, aminotriazole, led to significant increase in hydrogen peroxide concentration and *cAPX* transcription in rice cell cultures. Under excess light stress, Arabidopsis *cAPX* (*APX2*) expression followed an increase in hydrogen peroxide, and ABA appears to augment the role of hydrogen peroxide in initiating *APX2* expression. Furthermore, leaf transpiration rate also increased prior to *APX2* expression, suggesting that water status may also be involved in the signalling pathway [95]. This suggests that cytosolic APX is regulated by the hydrogen peroxide level within cells. Hydrogen peroxide accumulation induced by the signal transduction plays an important role in *cAPX* gene expression under stress conditions.

More research work has been focused on APX since APX plays an important role in clearing ROS. Arabidopsis cytoplasmic *apx1* mutation resulted in hydrogen peroxide accumulation under normal growth conditions [96]. Under light stress conditions, the *apx1* mutant showed growth retardation, altered stomata response, and an increasing induction of heat shock protein. The *tAPX* mutation in wheat resulted in lowered enzyme activity 40 % less than that of wild-type control, with a significant reduction in photosynthetic efficiency [97].

Cotton plants overexpressing *APX* showed improved chilling-stress tolerance [98]. Transgenic tobacco plants expressing antisense RNA for cytosolic APX show increased susceptibility to ozone injury [99]. Overexpressing Arabidopsis peroxisome APX gene (*APX3*) in tobacco resulted in improved protective capability against oxygen adversity [100] and increased seed number under water deficit [101]. Simultaneous expression of *SOD* and *APX* genes led to increasing tolerance to the herbicide paraquat-mediated oxidative stress in tobacco chloroplast [102], and resulted in even wider range of abiotic stress tolerance in tall fescue plants [103]. Overexpressing Arabidopsis *tAPX* gene resulted in increasing tolerance to the oxidative stress caused by paraquat [104]. However, the paraquat sensitivity of transgenic tobacco overproducing *APX* did not differ significantly from that of wild-type control plants [105].

On the other hand, transgenic tobacco plants overexpressing the *SOD* gene also exhibited a 3- to 4-fold increase in APX activity and had a corresponding increase in levels of APX mRNA, while dehydroascorbate reductase activities were not altered. The enhancement of the active oxygen-scavenging system that leads to increased oxidative stress protection in SOD overexpressing plants could result not only from increased SOD levels but from the combined increases in APX activity, providing the synergism effect between antioxidants of SOD and APX [106].

Curiously, the antisense transgenic tobacco with simultaneous inhibiting of the *APX* and *CAT* expression were less sensitive to oxygen adversity than those with separate inhibition of *APX* or *CAT* [107], indicating a compensation mechanism may happen when both *APX* and *CAT* expression are inhibited in plants. Siminlar compensation effect was also observed in *APX* co-suppressed tobacco cells. Ishikawa et al. [108] investigated the expression of Arabidopsis cytoplasmic APX gene (*APX1*) in tobacco cell line BY-2. In the transgenic cell line cAPX-S2 and cAPX-S3, co-suppression resulted in 50 and 70 % decreasing of APX activity, respectively, with significant rising of ROS levels. Interestingly, the two *APX* co-suppression cell lines exhibited resistance to high temperature and salt stress. Suppression subtractive hybridization showed that several high-temperature and salt stress-inducible genes were up-regulated in the cAPX-S3 cell line, indicating that decreasing APX activity may cause destruction of intracellular redox homeostasis, and activate the MAP kinase cascade.

The *APX* gene is also down-regulated to reduce ascorbate oxidation and promote ascorbate accumulation in plants. A mitochondrial *APX* was suppressed by RNA interference in tomato, resulting in lowered APX enzyme activity in both cytoplasm (0.4- to 0.8-fold) and mitochondrion (0.3- to 0.5-fold). The down-regulated *mitAPX* expression and APX enzyme activity lead to 1.4- to 2.2-fold increase in ascorbate content in tomato fruits as compared with wild-type control [93].

7.11   Dehydroascorbate Reductase (DHAR: EC.1.8.5.1)

Dehydroascorbate reductase (DHAR) as well as monodehydroascorbate reductase (MDHAR) is the enzyme responsible for ascorbate regeneration, catalyzing the conversion of oxidized state of dehydroascorbate to reduced state of ascorbate with glutathione as the electron donor.

DHAR, a monomeric thiol enzyme, has been purified from spinach [109, 110] and potato [111]. Convincing evidence shows that DHAR is widely present in various tissues of the plants [77]. Two isozymes of DHAR are identified, one in the cytoplasm, such as rice DHAR, and another plastid enzyme, such as DHAR purified from spinach leaves [109]. The catalytic mechanism of the spinach chloroplast

DHAR was investigated. Site-directed mutagenesis indicated that C23 in DHAR is indispensable for the reduction of dehydroascorbate [112].

The DHAR gene was firstly cloned from rice, and the recombinant protein expressed in *E. coli* confirmed that the gene encoded a functional DHAR enzyme [113]. Thereafter, the *DHAR* genes have been successively cloned from spinach, Arabidopsis, wheat, acerola, tobacco, tomato, potato and other plant species [108, 19, 114–116].

Transgenic approach was utilized to analyze the function of DHAR. Overexpression of wheat *DHAR* gene in tobacco and maize resulted in 32-fold and 100-fold increasing of DHAR enzyme activity in transgenic tobacco and maize, respectively. The *DHAR* overexpression led to the 2-4-fold increasing of ascorbate content in maize leaves and grains [114]. The content of reduced state of ascorbic acid as well as glutathione was also increased in the transgenic tobacco and maize, indicating that the *DHAR* overexpression can promote the ascorbate recycling and accumulation in plants.

It is reported the DHAR activity increases in response to various stress conditions and regulation of DHAR enzyme results in altered response to stress. The influence of redox state of ascorbate on the guard cell signaling and stomatal movement was investigated in the transgenic tobacco overexpressing *DHAR* [117]. The *DHAR* overexpressing resulted in increasing level of the reduced state of ascorbic acid in the guard cells, increasing stomatal openness and opening proportion, decreasing hydrogen peroxide accumulation, decreasing sensitivity of guard cells to the signals of hydrogen peroxide and ABA, and thus more water loss of plant in response to drought stresses. It could be argued that the reduced state of ascorbate plays an important role in the control of hydrogen peroxide-mediated stomatal closure. Chen and Gallie also found that transgenic plants overexpressing *DHAR* exhibited improved resistance to ozone [118].

7.12 Monodehydroascorbate Reductase (MDHAR: EC.1.6.5.4)

Monodehydroascorbate reductase (MDHAR) catalyzes the conversion from the oxidized state of monodehydroascorbate to reduced state of ascorbic acid using NADH or NADPH as an electron donor. As the enzyme for ascorbate regeneration, MDHAR plays an important role for maintaining the antioxidant properties of ascorbate.

In plant cells, MDHAR is not only present in the chloroplast [107], but also distributed in the cytoplasm [119, 120], mitochondrion [121], peroxisome [122], glyoxysome, and even on the plasmalemma. Several MDHAR isozymes from cytoplasm, mitochondrion and chloroplast have been purified from cucumber [109], potato [121], soybean, and spinach [123]. The cucumber MDHAR is a monomer. It contains 1 mol of FAD/mol of enzyme which is reduced by NAD(P)H and reoxidized by monodehydroascorbate. The MDHAR enzyme has an exposed thiol group whose blockage with thiol reagents inhibits the electron transfer from NAD(P)H to the enzyme FAD. Both NADH and NADPH served as electron donors [109].

The full-length cDNA of *MDHAR* has been isolated from cucumber [124], pea [125], tomato [119, 19], Chinese cabbage [120] and other plant species. In addition, a peroxisome *MDHAR* gene was isolated from the pea genome. The pea *MDHAR* is 3,770 bp in full length of with nine exons and eight introns [122].

MDHAR enzyme activity varies in different cell growth stages and plant tissues, leading to the different ascorbate/monodehydroascorbate ratios. In meristematic cells MDHAR activity is very high, and consequently a large amount of monodehydroascorbate is reduced to ascorbate; in expanding cells, however, the MDHAR activity is relatively lower, and therefore the accumulated monodehydroascorbate leads to high level of dehydroascorbate [126]. These data show MDHAR enzyme activity as well as ratio of ascorbate/monodehydroascorbate is closely related with the cell growth. In addition, the MDHAR-mediated dehydroascorbate reduction is supposed to play an important role during the early stage of seed germination, as the dry seed only contains dehydroascorbate without ascorbate [126]. *MDHAR* gene expression and MDHAR enzyme activity in roots were the higher than those in other tissues tested in tomato. The ascorbate content in tomato roots, however, was the lowest, suggesting the final ascorbate level results from the combined action of synthesis, oxidization, and recycling [119].

The *MDHAR* gene expression increased rapidly in various tissues after stress. The cabbage *MDHAR* gene was constitutively expressed in all tissues tested, and up-regulated in response to hydrogen peroxide, salicylic acid, paraquat, and ozone, indicating MDHAR plays an important role in clearing ROS possibly through a transcriptional regulation machinery [120].

Overexpressing *MDHAR* can be utilized to enhance the ascorbate accumulation in plants. Ectopic expressing of Acerola *MDHAR* gene in tobacco resulted in higher MDHAR activity and greater accumulation of ascorbate in transgenic plants [116]. The transgenic tomato overexpressing *MDHAR* exhibited up to 3-fold increase in MDHAR enzyme activity and 21 % increase in ascorbate content compared to wild-type control (unpublished data).

7.13 Myoinositol Oxygenase (MIOX: EC 1.13.99.1)

Myoinositol oxygenase (MIOX) is one of the potential enzymes for ascorbate biosynthesis in plant oxidizing the myoinositol into D-glucuronate. MIOX is an enzyme containing non-heme iron and catalyzes a four-electron oxidation with the transfer of only one atom of oxygen into the product D-glucuronate using iron as its cofactor.

MIOX has not been widely investigated in plants. Transgenic research about *MIOX* has been done in Arabidopsis and tomato, and its expression patterns have been investigated in Arabidopsis, tomato and rice. However, its role in ascorbate biosynthesis remains controversial.

The early experiments of feeding ripening strawberries and celery leaves with radioactively labeled myoinositol did not produce radioactively labeled ascorbate. On the contrary, the majority of radioactive labels were concentrated in the pectic polysaccharide [127]. Completion of the Arabidopsis genome sequencing project allows researchers to find plant homologous gene with the pig myoinositol oxygenase gene. Genome sequence data mining discovered four open reading frames (ORF) homologous to pig myoinositol oxygenase gene on Arabidopsis chromosome 1, 2, 4 and 5, respectively. Lorence et al. [128] proposed the myoinositol pathway of plant ascorbate biosynthesis after they cloned the myoinositol oxygenase gene from Arabidopsis. Overexpression of *MIOX4* gene resulted in 2-3-fold increase in ascorbate content [128]. Overexpressing strawberry *MIOX* gene in the transgenic tomato also resulted in increasing ascorbate content in leaves under myoinositol feeding (unpublished).

On the other hand, the ascorbate content in the leaves of the Arabidopsis *MIOX2* or *MIOX5* mutants was not altered by the gene mutation, indicating that these genes of *MIOX* family may be more involved in the biosynthesis of nucleotide sugar precursors for cell-wall matrix polysaccharides than in ascorbate synthesis [129].

Furthermore, overexpression of *MIOX* resulted in >30 fold-increase in its enzyme activity, but did not affect the ascorbate accumulation. Overexpression of *MIOX* led to less accumulation of myoinositol compared to the wild-type control under myoinositol feeding, possibly because of the gene overexpressing and increasing MIOX enzyme activity. The ascorbate remained the same level between the *MIOX*-overexpressing lines and the wild-type control. Even under the low light or salt stress when the ascorbate content increased 4 folds, no difference was observed in the ascorbate biosynthesis between *MIOX*-overexpressing lines and the wild-type control. The exogenous feeding with myoinositol did not affect the ascorbate content in both transgenic plants and wild-type control, suggesting that the MIOX catalyzing may not be a key step for ascorbate synthesis in plants [130]. This is inconsistent with the previous results of Lorence et al. [128]. More work should be done about the MIOX pathway in ascorbate synthesis.

Therefore, there are two statements on the role of *MIOX* in the ascorbate biosynthesis pathway. The first point of view argued that overexpression of *MIOX* can improve the ascorbate levels in *Arabidopsis thaliana*. Overexpression of *miox4* resulted in 2–3-fold increasing in the ascorbate content in leaves [128]. The other point of view, however, considers the MIOX catalyzing as a negligible step in ascorbate synthesis. Overexpression of *MIOX* led to 30-fold increase in the MIOX enzyme activity, but did not affect the ascorbate content even under various stress conditions [130]. Obviously, these two results are contradictory. Overexpressing strawberry *MIOX* gene in tomato improved the MIOX enzyme activity, but did not affect the ascorbate accumulation in tomato significantly under normal conditions. However, when fed with exogenous myoinositol, the *MIOX*-overexpressing transgenic tomato showed higher ascorbate content in both leaves and fruits compared to the wild-type control unpublished data. Therefore, we are more inclined to the first statement of MIOX role.

MIOX gene expression shows species-specific pattern. Tomato *MIOX* gene was up-regulated by high temperature, while no gene expression was observed under polyethylene glycol (PEG) stress. The tomato *MIOX* gene showed highest expression in the stems, young flowers and young fruits, and low expression levels in young leaves and mature leaves (unpublished data). In the Arabidopsis, however, the *MIOX4* was highly expressed in leaves and flowers with very low expression in stems, indicating different regulating pattern of *MIOX* in different plant species [128].

7.14 Alternative Pathway and Enzyme for Ascorbate Biosynthesis and Metabolism

In addition to the genes and enzymes mentioned above, several other enzymes and optional pathways were also investigated in the context of plant ascorbate biosynthesis and metabolism.

7.14.1 L-Gulono-1,4-Lactone Oxidase (GLOase: EC 1.1.3.8)

A successful attempt to improve the ascorbate content was achieved through genetic engineering L-gulono-1,4-lactone oxidase gene (GLOase). The gene encoding L-gulono-1,4-lactone oxidase (the last step enzyme of ascorbate synthesis in animals) in rat was transferred and expressed in the lettuce and tobacco, resulting a up to 7-fold increase in ascorbate content in transgenic plants [131]. The same research group reported that the heterologous expression of rate *GLOase* gene in Arabidopsis could restore ascorbate accumulation in the five Arabidopsis ascorbate defective mutants. And the wild-type Arabidopsis plants transformed with the *GLOase* showed up to a 2-fold increase in ascorbate content in leaves compared to untransformed control [132]. Transgenic potato overexpressing L-gulono-γ-lactone oxidase gene (*GLOase*) from rat showed enhanced ascorbate accumulation and better survival under various abiotic stresses caused by methyl viologen (MV), NaCl or mannitol, respectively [133].

These results demonstrate that basal levels of ascorbate in plants can be significantly increased by expressing a single gene from the animal pathway, and that alternative biosynthetic pathway for ascorbate may work in plants.

7.14.2 D-Galacturonate Reductase (GalUR: EC 1.1.1.203)

D-galacturonate reductase (GalUR) catalyses the conversion of D-galacturonic acid to L-galactonic acid. L-galactonic acid is utilized to synthesize L-ascorbic acid through the intermediate of L-galactono-1,4-lactone by GLDH catalyzing. The substrate of the GalUR enzyme, D-galacturonic acid, is the main component of pectin. The *GalUR* gene was isolated and identified from strawberry, and correlation between *GalUR* expression level and ascorbate content in strawberry fruit was observed [134]. Overexpression of *GalUR* gene in Arabidopsis resulted in 2-3-fold increase in the ascorbate content compared to the wild-type control [134]. Ectopic expression of strawberry *GalUR* in tomato significantly increased the total ascorbate content in fruits, with the maximum increase of 65 % compared with wild-type plants. This demonstrates the feasibility to increase tomato ascorbate content by overexpressing *FaGalUR*. Also, it provides evidence that D-galacturonic acid pathway may be an alternative mechanism for ascorbate biosynthesis in tomato (unpublished data).

7.15 The Multi-Gene Family Involved in Ascorbate Biosynthesis and Metabolism

Ascorbate metabolism related enzymes, such as the APX, DHAR, MDHAR, and GME normally are encoded by genes of multi-gene family.

APX is an important enzyme for clearing hydrogen peroxide especially those in the chloroplast. It is distributed in various cell compartments such as the cytoplasm, chloroplast, microbody and mitochondria, and the APX in chloroplasts includes stromal APX and thylakoid-membrane-bound APX. The APX isozymes show diversity in their composition, structure, substrate specificity, affinity, and purifying stability [108, 81]. The multi-gene family for APX may come from the alternative splicing of the pre-mRNA [85].

Two cytoplasmic APX genes were cloned in tomato, namely APX20 [89] and LecAPX2 [19], with 88 % homology between the two amino acid sequences. LecAPX2 showed a 92 % homology with potato cAPX in amino acid sequence. Amino acid sequence of the LecAPX2 contains signals of heme peroxidase and plant APX family. According to comparison of tomato *cAPX* genes with the counterparts from other plant species, it is clear that there are at least two different cytosolic *APX* genes in tomato. Higher plants have two or more cAPXs demonstrating that cAPX isozyme is encoded by a multi-gene family. Three different *cAPX* genes were identified in Arabidopsis [135]. Two *cAPX* cDNA were isolated from soybean leaves by screening the cDNA library [136]. In rice, two *cAPX* genes (*OsAPX1*, and *OsAPX2*) were also cloned and characterized [91].

DHAR, a key enzyme for ascorbate metabolic pathway promoting the regeneration of ascorbate, is also encoded by multi-gene family. Thus far two DHAR isozymes are identified: one in the cytoplasm, such as rice DHAR, another in plastid, such as chloroplast DHAR purified from spinach [109]. The full-length cDNAs of two *DHAR* genes (*DHAR1* and *DHAR2*) were cloned from tomato [19]. Subcellular localization prediction and homology analysis shows that the DHAR1 belongs to the cytoplasmic enzyme, and DHAR2 belongs to the plastid type. There are at least two different DHAR isozymes in tomato, and the existence of other DHAR isozyme is not excluded.

Two *GME* genes (*SlGME1* and *SlGME2*) involved in ascorbate Smirnoff biosynthesis pathway were cloned from tomato under the accession number GQ150164 and GQ150165, respectively. They both contain a complete open reading frame of 1131 bp (six exons and five introns) encoding 376 amino acids. The nucleotide sequences of the two *SlGME* genes share 79 % identity, whereas the two putative amino acid sequences have 92 % homology. The fusion protein of *SlGME1* and *SlGME2* with *GFP* were transiently expressed in onion epidermal cells, showing that the fluorescence of both fusion proteins of SlGME1::GFP and SlGME2::GFP are clearly visible in the cytoplasm. *SlGME1* and *SlGME2* were constitutively expressed in roots, stems, leaves, flowers and fruits, whereas different expression abundance were found among different tissues [32].

The factor that most of ascorbate biosynthesis and metabolism-related enzymes in plants are encoded by multi-gene family may be attributed to the highly redundant and flexible plant genome. The multi-gene of the enzymes may collectively contribute to fulfilling the biological process. The Arabidopsis ascorbate defective mutants are able to survive despite of the sharp decreasing of ascorbate in plants. The still remaining ascorbate level in the ascorbate defective mutants may result from the multi-gene family, which can compensate for the dysfunctions of gene member.

This multi-gene family can be beneficial for plant growth and response to environment by spatial and temporal-specific expression. The expression levels of ascorbate metabolism-related genes vary among different tissues and different developmental stages. The tomato expression database analysis showed that among the 14 genes related to ascorbate biosynthesis and metabolism in tomato, 9 genes were expressed at a tissue-specific pattern. The other 5 genes were constitutively expressed in the roots, leaves, flowers and fruits, with the varying expression abundance at different developmental stages of the same tissue. The ascorbate metabolism related-enzyme encoding genes are differentially expressed in response to different environmental conditions, especially under various stresses. In the early growth stage after germination of Japanese radish, the expression of *cAPX* gene was rapidly induced in roots and regulated by light [137]. Exogenous hydrogen peroxide can also alter the transcriptional abundance of *cAPX* in cultured soybean cells [138]. Treatment with hydrogen peroxide, salicylic acid, paraquat and ozone resulted in elevated expression of *MDHAR* gene in cabbage [120]. Therefore, the multi-gene family for ascorbate metabolism related enzymes may help plant adapt to environments and stresses.

On the other hand, the multi-gene family of ascorbate metabolism related enzymes may make it more difficult to decipher the function of each member of gene family by means of RNAi or antisense technology.

References

1. Negrotto D, Jolley M, Beer S, Wenck AR, Hansen G (2000) The use of phosphomannose-isomerase as a selectable marker to recover transgenic maize plants (*Zea mays* L.) via *Agrobacterium* transformation. Plant Cell Rep 19:798–803
2. Briza J, Ruzickova N, Niedermeierova H, Dusbaskova J, Vlasak J (2010) Phosphomannose isomerase gene for selection in lettuce (*Lactuca sativa* L.) transformation. Acta Biochim Pol 57:63–68
3. Wallbraun M, Sonntag K, Eisenhauer C, Krzcal G, Wang YP (2009) Phosphomannose isomerase (PMI) gene as a selectable marker for *Agrobacterium*-mediated transformation of rapeseed. Plant Cell Tiss Org Cult 99:345–351
4. He ZQ, Fu YP, Si HM, Hu GC, Zhang SH, Yu YH, Sun ZX (2004) Phosphomannose-isomerase (PMI) gene as a selectable marker for rice transformation via *Agrobacterium*. Plant Sci 166:17–22
5. Briza J, Pavingerova D, Prikrylova P, Gazdova J, Vlasak J, Niedermeierova H (2008) Use of phosphomannose isomerase-based selection system for *Agrobacterium*-mediated transformation of tomato and potato. Biol Plant 52:453–461
6. Gadaleta A, Giancaspro A, Blechl A, Blanco A (2006) Phosphomannose isomerase, PMI, as a selectable marker gene for durum wheat transformation. J Cereal Sci 43:31–37
7. Degenhardt J, Poppe A, Montag J, Szankowski I (2006) The use of the phosphomannose-isomerase/mannose selection system to recover transgenic apple plants. Plant Cell Rep 25:1149–1156
8. Min BW, Cho YN, Song MJ, Noh TK, Kim BK, Chae WK, Park YS, Choi YD, Harn CH (2007) Successful genetic transformation of Chinese cabbage using phosphomannose isomerase as a selection marker. Plant Cell Rep 26:337–344
9. Lamblin F, Aime A, Hano C, Roussy I, Domon JM, Van Droogenbroeck B, Laine E (2007) The use of the phosphomannose isomerase gene as alternative selectable marker for *Agrobacterium*-mediated transformation of flax (*Linum usitatissimum*). Plant Cell Rep 26:765–772
10. Maruta T, Yonemitsu M, Yabuta Y, Tamoi M, Ishikawa T, Shigeoka S (2008) Arabidopsis phosphomannose isomerase 1, but not phosphomannose isomerase 2, is essential for ascorbic acid biosynthesis. J Biol Chem 283:28842–28851
11. Oesterhelt C, Schnarrenberger C, Gross W (1996) Phosphomannomutase and phosphoglucomutase in the red alga *Galdieria sulphuraria*. Plant Sci 121:19–27
12. Popova TN, Matasova LV, Lapot'ko AA (1998) Purification, separation and characterization of phosphoglucomutase and phosphomannomutase from maize leaves. Biochem Mol Biol Int 46:461–470
13. Oesterhelt C, Schnarrenberger C, Gross W (1997) The reaction mechanism of phosphomannomutase in plants. FEBS Lett 401:35–37
14. Qian WQ, Yu CM, Qin HJ, Liu X, Zhang AM, Johansen IE, Wang DW (2007) Molecular and functional analysis of phosphomannomutase (PMM) from higher plants and genetic evidence for the involvement of PMM in ascorbic acid biosynthesis in Arabidopsis and *Nicotiana benthamiana*. Plant J 49:399–413
15. Badejo AA, Eltelib HA, Fukunaga K, Fujikawa Y, Esaka M (2009) Increase in ascorbate content of transgenic tobacco plants overexpressing the acerola (*Malpighia glabra*) phosphomannomutase gene. Plant Cell Physiol 50:423–428
16. Wheeler GL, Jones MA, Smirnoff N (1998) The biosynthetic pathway of vitamin C in higher plants. Nature 393:365–369
17. Conklin PL, Norris SR, Wheeler GL, Williams EH, Smirnoff N, Last RL (1999) Genetic evidence for the role of GDP-mannose in plant ascorbic acid (vitamin C) biosynthesis. Proc Nat Acad Sci USA 96:4198–4203

18. Keller R, Springer F, Renz A, Kossmann J (1999) Antisense inhibition of the GDP-mannose pyrophosphorylase reduces the ascorbate content in transgenic plants leading to developmental changes during senescence. Plant J 19:131–141

19. Zou LP, Li HX, Ouyang B, Zhang JH, Ye ZB (2006) Cloning and mapping of genes involved in tomato ascorbic acid biosynthesis and metabolism. Plant Sci 170:120–127

20. Badejo AA, Jeong ST, Goto-Yamamoto N, Esaka M (2007) Cloning and expression of GDP-D-mannose pyrophosphorylase gene and ascorbic acid content of acerola (*Malpighia glabra* L.) fruit at ripening stages. Plant Physiol Biochem 45:665–672

21. Conklin PL, Williams EH, Last RL (1996) Environmental stress sensitivity of an ascorbic acid-deficient Arabidopsis mutant. Proc Nat Acad Sci USA 93:9970–9974

22. Badejo AA, Tanaka N, Esaka M (2008) Analysis of GDP-D-mannose pyrophosphorylase gene promoter from acerola (*Malpighia glabra*) and increase in ascorbate content of transgenic tobacco expressing the acerola gene. Plant Cell Physiol 49:126–132

23. Lukowitz W, Nickle TC, Meinke DW, Last RL, Conklin PL, Somerville CR (2001) Arabidopsis cyt1 mutants are deficient in a mannose-1-phosphate guanylyltransferase and point to a requirement of N-linked glycosylation for cellulose biosynthesis. Proc Nat Acad Sci USA 98:2262–2267

24. Veljovic-Jovanovic SD, Pignocchi C, Noctor G, Foyer CH (2001) Low ascorbic acid in the *vtc-1* mutant of Arabidopsis is associated with decreased growth and intracellular redistribution of the antioxidant system. Plant Physiol 127:426–435

25. Qin C, Qian WQ, Wang WF, Wu Y, Yu CM, Jiang XH, Wang DW, Wu P (2008) GDP-mannose pyrophosphorylase is a genetic determinant of ammonium sensitivity in *Arabidopsis thaliana*. Proc Nat Acad Sci USA 105:18308–18313

26. Barth C, Gouzd ZA, Steele HP, Imperio RM (2010) A mutation in GDP-mannose pyrophosphorylase causes conditional hypersensitivity to ammonium, resulting in Arabidopsis root growth inhibition, altered ammonium metabolism, and hormone homeostasis. J Exp Bot 61:379–394

27. Li Q, Li BH, Kronzucker HJ, Shi WM (2010) Root growth inhibition by NH₄⁺ in Arabidopsis is mediated by the root tip and is linked to NH₄⁺ efflux and GMPase activity. Plant Cell Environ 33:1529–1542

28. Wolucka BA, Persiau G, Van Doorsselaere J, Davey MW, Demol H, Vandekerckhove J, Van Montagu M, Zabeau M, Boerjan W (2001) Partial purification and identification of GDP-mannose 3', 5'-epimerase of *Arabidopsis thaliana*, a key enzyme of the plant vitamin C pathway. Proc Nat Acad Sci USA 98:14843–14848

29. Wolucka BA, Van Montagu M (2003) GDP-mannose 3', 5'-epimerase forms GDP-L-gulose, a putative intermediate for the de novo biosynthesis of vitamin C in plants. J Biol Chem 278:47483–47490

30. Valpuesta V, Botella MA (2004) Biosynthesis of L-ascorbic acid in plants: new pathways for an old antioxidant. Trends Plant Sci 9:573–577

31. Watanabe K, Suzuki K, Kitamura S (2006) Characterization of a GDP-D-mannose 3', 5'-epimerase from rice. Phytochemistry 67:338–346

32. Zhang CJ, Liu JX, Zhang YY, Cai XF, Gong PJ, Zhang JH, Wang TT, Li HX, Ye ZB (2011) Overexpression of *SlGMEs* leads to ascorbate accumulation with enhanced oxidative stress, cold, and salt tolerance in tomato. Plant Cell Rep 30:389–398

33. Stevens R, Buret M, Duffe P, Garchery C, Baldet P, Rothan C, Causse M (2007) Candidate genes and quantitative trait loci affecting fruit ascorbic acid content in three tomato populations. Plant Physiol 143:1943–1953

34. Gilbert L, Alhagdow M, Nunes-Nesi A, Quemener B, Guillon F, Bouchet B, Faurobert M, Gouble B, Page D, Garcia V, Petit J, Stevens R, Causse M, Fernie AR, Lahaye M, Rothan C, Baldet P (2009) GDP-D-mannose 3,5-epimerase (GME) plays a key role at the intersection of ascorbate and non-cellulosic cell-wall biosynthesis in tomato. Plant J 60:499–508

35. Conklin PL, Saracco SA, Norris SR, Last RL (2000) Identification of ascorbic acid-deficient *Arabidopsis thaliana* mutants. Genetics 154:847–856

36. Laing WA, Wright MA, Cooney J, Bulley SM (2007) The missing step of the L-galactose pathway of ascorbate biosynthesis in plants, an L-galactose guanyltransferase, increases ascorbate content. Proc Nat Acad Sci USA 104:9534–9539

37. Linster CL, Gomez TA, Christensen KC, Adler LN, Young BD, Brenner C, Clarke SG (2007) Arabidopsis VTC2 encodes a GDP-L-galactose phosphorylase, the last unknown enzyme in the Smirnoff-Wheeler pathway to ascorbic acid in plants. J Biol Chem 282:18879–18885

38. Dowdle J, Ishikawa T, Gatzek S, Rolinski S, Smirnoff N (2007) Two genes in *Arabidopsis thaliana* encoding GDP-L-galactose phosphorylase are required for ascorbate biosynthesis and seedling viability. Plant J 52:673–689

39. Muller-Moule P (2008) An expression analysis of the ascorbate biosynthesis enzyme VTC2. Plant Mol Biol 68:31–41

40. Bulley SM, Rassam M, Hoser D, Otto W, Schunemann N, Wright M, MacRae E, Gleave A, Laing W (2009) Gene expression studies in kiwifruit and gene over-expression in Arabidopsis indicates that GDP-L-galactose guanyltransferase is a major control point of vitamin C biosynthesis. J Exp Bot 60:765–778

41. Linster CL, Clarke SG (2008) L-Ascorbate biosynthesis in higher plants: the role of VTC2. Trends Plant Sci 13:567–573

42. Laing WA, Frearson N, Bulley S, MacRae E (2004) Kiwifruit L-galactose dehydrogenase: molecular, biochemical and physiological aspects of the enzyme. Funct Plant Biol 31:1015–1025

43. Laing WA, Bulley S, Wright M, Cooney J, Jensen D, Barraclough D, MacRae E (2004) A highly specific L-galactose-1-phosphate phosphatase on the path to ascorbate biosynthesis. Proc Nat Acad Sci USA 101:16976–16981

44. Conklin PL, Gatzek S, Wheeler GL, Dowdle J, Raymond MJ, Rolinski S, Isupov M, Littlechild JA, Smirnoff N (2006) *Arabidopsis thaliana* VTC4 encodes L-galactose-1-P phosphatase, a plant ascorbic acid biosynthetic enzyme. J Biol Chem 281:15662–15670

45. Ioannidi E, Kalamaki MS, Engineer C, Pateraki I, Alexandrou D, Mellidou I, Giovannonni J, Kanellis AK (2009) Expression profiling of ascorbic acid-related genes during tomato fruit development and ripening and in response to stress conditions. J Exp Bot 60:663–678

46. Torabinejad J, Donahue JL, Gunesekera BN, Allen-Daniels MJ, Gillaspy GE (2009) VTC4 is a bifunctional enzyme that affects myoinositol and ascorbate biosynthesis in plants. Plant Physiol 150:951–961

47. Gatzek S (2002) Antisense suppression of L-galactose dehydrogenase in *Arabidopsis thaliana* provides evidence for its role in ascorbate-synthesis and reveals light modulated L-galactose synthesis. Plant J 31:553–553

48. Mieda T, Yabuta Y, Rapolu M, Motoki T, Takeda T, Yoshimura K, Ishikawa T, Shigeoka S (2004) Feedback inhibition of spinach L-galactose dehydrogenase by L-ascorbate. Plant Cell Physiol 45:1271–1279

49. Smirnoff N (2000) Ascorbic acid: metabolism and functions of a multi-facetted molecule. Curr Opin Plant Biol 3:229–235

50. Oba K, Ishikawa S, Nishikawa M, Mizuno H, Yamamoto T (1995) Purification and properties of L-galactono-gamma-lactone dehydrogenase, a key enzyme for ascorbic acid biosynthesis, from sweet potato roots. J Biochem 117:120–124

51. Imai T, Karita S, Shiratori G, Hattori M, Nunome T, Oba K, Hirai M (1998) L-galactono-γ-lactone dehydrogenase from sweet potato: purification and cDNA sequence analysis. Plant Cell Physiol 39:1350–1358

52. Ostergaard J, Persiau G, Davey MW, Bauw G, VanMontagu M (1997) Isolation of a cDNA coding for L-galactono-gamma-lactone dehydrogenase, an enzyme involved in the biosynthesis of ascorbic acid in plants. Purification, characterization, cDNA cloning, and expression in yeast. J Biol Chem 272:30009–30016

53. Mutsuda M, Ishikawa T, Takeda T, Shigeoka S (1995) Subcellular-localization and properties of L-galactono-γ-lactone dehydrogenase in spinach leaves. Biosci Biotechnol Biochem 59:1983–1984

54. Yabuta Y, Yoshimura K, Takeda T, Shigeoka S (2000) Molecular characterization of tobacco mitochondrial L-galactono-gamma-lactone dehydrogenase and its expression in *Escherichia coli*. Plant Cell Physiol 41:666–675

55. do Nascimento JRO, Higuchi BK, Gomez MLPA, Oshiro RA, Lajolo FM (2005) L-ascorbate biosynthesis in strawberries: L-galactono-1,4-lactone dehydrogenase expression during fruit development and ripening. Postharvest Biol Technol 38:34–42

56. Pateraki I, Sanmartin M, Kalamaki MS, Gerasopoulos B, Kanellis AK (2004) Molecular characterization and expression studies during melon fruit development and ripening of L-galactono-1,4-lactone dehydrogenase. J Exp Bot 55:1623–1633

57. Zhang YY, Li HX, Shu WB, Zhang CJ, Zhang W, Ye ZB (2011) Suppressed expression of ascorbate oxidase gene promotes ascorbic acid accumulation in tomato fruit. Plant Mol Biol Rep 29:638–645

58. Leferink NGH, van den Berg WAM, van Berkel WJH (2008) L-Galactono-γ-lactone dehydrogenase from *Arabidopsis thaliana*, a flavoprotein involved in vitamin C biosynthesis. FEBS J 275:713–726.

59. Arrigoni O, DeGara L, Paciolla C, Evidente A, dePinto MC, Liso R (1997) Lycorine: a powerful inhibitor of L-galactono-gamma-lactone dehydrogenase activity. J Plant Physiol 150:362–364

60. Leferink NGH, van den Berg, WAM, van Berkel WJH (2008) L-Galactono-γ-lactone dehydrogenase from *Arabidopsis thaliana*, a flavoprotein involved in vitamin C biosynthesis. FEBS J 275:713–726.

61. Tabata K, Takaoka T, Esaka M (2002) Gene expression of ascorbic acid-related enzymes in tobacco. Phytochemistry 61:631–635

62. Tabata K, Oba K, Suzuki K, Esaka M (2001) Generation and properties of ascorbic acid-deficient transgenic tobacco cells expressing antisense RNA for L-galactono-1,4-lactone dehydrogenase. Plant Journal 27:139–148

63. Tokunaga T, Miyahara K, Tabata K, Esaka M (2005) Generation and properties of ascorbic acid-overproducing transgenic tobacco cells expressing sense RNA for L-galactono-1, 4-lactone dehydrogenase. Planta 220:854–863

64. Imai T, Niwa M, Ban Y, Hirai M, Oba K, Moriguchi T (2009) Importance of the L-galactonolactone pool for enhancing the ascorbate content revealed by L-galactonolactone dehydrogenase-overexpressing tobacco plants. Plant Cell Tiss Org Cult 96:105–112

65. Alhagdow M, Mounet F, Gilbert L, Nunes-Nesi A, Garcia V, Just D, Petit J, Beauvoit B, Fernie AR, Rothan C, Baldet P (2007) Silencing of the mitochondrial ascorbate synthesizing enzyme L-galactono-1,4-lactone dehydrogenase affects plant and fruit development in tomato. Plant Physiol 145:1408–1422

66. Pineau B, Layoune O, Danon A, De Paepe R (2008) L-Galactono-1,4-lactone dehydrogenase is required for the accumulation of plant respiratory complex I. J Biol Chem 283:32500–32505

67. Smirnoff N (1996) The function and metabolism of ascorbic acid in plants. Ann Bot 78:661–669

68. Esaka M, Hattori T, Fujisawa K, Sakajo S, Asahi T (1990) Molecular cloning and nucleotide sequence of full-length cDNA for ascorbate oxidase from cultured pumpkin cells. Eur J Biochem 191:537–541

69. Moser O, Kanellis AK (1994) Ascorbate oxidase of *Cucumis melo* L. var. reticulatus: purification, characterization and antibody production. J Exp Bot 45:717–724

70. Ohkawa J, Okada N, Shinmyo A, Takano M (1989) Primary structure of cucumber (*Cucumis sativus*) ascorbate oxidase deduced from cDNA sequence: homology with blue copper proteins and tissue-specific expression. Proc Nat Acad Sci USA 86:1239–1243

71. Ohkawa J, Ohya T, Ito T, Nozawa H, Nishi Y, Okada N, Yoshida K, Takano M, Shinmyo A (1994) Structure of the genomic dna encoding cucumber ascorbate oxidase and its expression in transgenic plants. Plant Cell Rep 13:481–488

72. Diallinas G, Pateraki I, Sanmartin M, Scossa A, Stilianou E, Panopoulos NJ, Kanellis AK (1997) Melon ascorbate oxidase: cloning of a multigene family, induction during fruit development and repression by wounding. Plant Mol Biol 34:759–770

73. Sanmartin M, Drogoudi PD, Lyons T, Pateraki I, Barnes J, Kanellis AK (2003) Over-expression of ascorbate oxidase in the apoplast of transgenic tobacco results in altered ascorbate and glutathione redox states and increased sensitivity to ozone. Planta 216:918–928

74. Kato N, Esaka M (2000) Expansion of transgenic tobacco protoplasts expressing pumpkin ascorbate oxidase is more rapid than that of wild-type protoplasts. Planta 210:1018–1022

75. Albani D, Sardana R, Robert LS, Altosaar I, Arnison PG, Fabijanski SF (1992) A *Brassica napus* gene family which shows sequence similarity to ascorbate oxidase is expressed in developing pollen. Molecular characterization and analysis of promoter activity in transgenic tobacco plants. Plant J 2:331–342

76. Pignocchi C, Foyer CH (2003) Apoplastic ascorbate metabolism and its role in the regulation of cell signalling. Curr Opin Plant Biol 6:379–389

77. Kato N, Esaka M (1999) Changes in ascorbate oxidase gene expression and ascorbate levels in cell division and cell elongation in tobacco cells. Physiol Plantarum 105:321–329

78. Pignocchi C, Fletcher JM, Wilkinson JE, Barnes JD, Foyer CH (2003) The function of ascorbate oxidase in tobacco. Plant Physiol 132:1631–1641

79. Yamamoto A, Bhuiyan NH, Waditee R, Tanaka Y, Esaka M, Oba K, Jagendorf AT, Takabe T (2005) Suppressed expression of the apoplastic ascorbate oxidase gene increases salt tolerance in tobacco and Arabidopsis plants. J Exp Bot 56:1785–1796

80. Felton GW, Summers CB (1993) Potential role of ascorbate oxidase as a plant defense protein against insect herbivory. J Chem Ecol 19:1553–1568

81. Shigeoka S, Ishikawa T, Tamoi M, Miyagawa Y, Takeda T, Yabuta Y, Yoshimura K (2002) Regulation and function of ascorbate peroxidase isoenzymes. J Exp Bot 53:1305–1319

82. Asada K (1992) Ascorbate peroxidase—a hydrogen peroxide-scavenging enzyme in plants. Physiol Plantarum 85:235–241

83. Amako K, Chen GX, Asada K (1994) Separate assays specific for ascorbate peroxidase and guaiacol peroxidase and for the chloroplastic and cytosolic isozymes of ascorbate peroxidase in plants. Plant Cell Physiol 35:497–504

84. Dabrowska G, Katai A, Goc A, Szechynska-Hebda M, Skrzypek E (2007) Characteristics of the plant ascorbate peroxidase family. Acta Biologica Cracoviensia Series Botanica 49:7–17

85. Yoshimura K, Yabuta Y, Ishikawa T, Shigeoka S (2002) Identification of a cis element for tissue-specific alternative splicing of chloroplast ascorbate peroxidase pre-mRNA in higher plants. J Biol Chem 277:40623–40632

86. Mittler R, Zilinskas BA (1992) Molecular-Cloning and characterization of a gene encoding pea cytosolic ascorbate peroxidase. J Biol Chem 267:21802–21807

87. Lopez F, Vansuyt G, CasseDelbart F, Fourcroy P (1996) Ascorbate peroxidase activity, not the mRNA level, is enhanced in salt-stressed *Raphanus sativus* plants. Physiol Plantarum 97:13–20

88. Santos M, Gousseau H, Lister C, Foyer C, Creissen G, Mullineaux P (1996) Cytosolic ascorbate peroxidase from *Arabidopsis thaliana* L is encoded by a small multigene family. Planta 198:64–69

89. Gadea J, Conejero V, Vera P (1999) Developmental regulation of a cytosolic ascorbate peroxidase gene from tomato plants. Mol Gen Genet 262:212–219

90. Kitajima S, Ueda M, Sano S, Miyake C, Kohchi T, Tomizawa K, Shigeoka S, Yokota A (2002) Stable form of ascorbate peroxidase from the red alga Galdieria partita similar to both chloroplastic and cytosolic isoforms of higher plants. Biosci Biotechnol Biochem 66:2367–2375

91. Agrawal GK, Jwa NS, Iwahashi H, Rakwal R (2003) Importance of ascorbate peroxidases OsAPX1 and OsAPX2 in the rice pathogen response pathways and growth and reproduction revealed by their transcriptional profiling. Gene 322:93–103

92. da Costa DS, Pereira CS, Teixeira J, Pereira S (2006) Isolation and characterisation of a cDNA encoding a novel cytosolic ascorbate peroxidase from potato plants (*Solanum tuberosum* L.). Acta Physiol Plant 28:41–47
93. Zhang YY, Li HX, Shu WB, Zhang CJ, Ye ZB (2011) RNA interference of a mitochondrial APX gene improves vitamin C accumulation in tomato fruit. Scientia Horticulturae 129:220–226
94. Desikan R, Mackerness SAH, Hancock JT, Neill SJ (2001) Regulation of the Arabidopsis transcriptome by oxidative stress. Plant Physiol 127:159–172
95. Fryer MJ, Ball L, Oxborough K, Karpinski S, Mullineaux PM, Baker NR (2003) Control of ascorbate peroxidase 2 expression by hydrogen peroxide and leaf water status during excess light stress reveals a functional organisation of Arabidopsis leaves. Plant J 33:691–705
96. Pnueli L, Liang H, Rozenberg M, Mittler R (2003) Growth suppression, altered stomatal responses, and augmented induction of heat shock proteins in cytosolic ascorbate peroxidase (APX1)-deficient Arabidopsis plants. Plant J 34:185–201
97. Danna CH, Bartoli CG, Sacco F, Ingala LR, Santa-Maria GE, Guiamet JJ, Ugalde RA (2003) Thylakoid-bound ascorbate peroxidase mutant exhibits impaired electron transport and photosynthetic activity. Plant Physiol 132:2116–2125
98. Payton P, Allen RD, Webb R, Holaday AS (1997) Chilling-stress tolerance of cotton plants over-expressing superoxide dismutase, ascorbate peroxidase, or glutathione reductase. Plant Physiol 114:593–593
99. Orvar BL, Ellis BE (1997) Transgenic tobacco plants expressing antisense RNA for cytosolic ascorbate peroxidase show increased susceptibility to ozone injury. Plant J 11:1297–1305
100. Wang WX, Vinocur B, Altman A (2003) Plant responses to drought, salinity and extreme temperatures: towards genetic engineering for stress tolerance. Planta 218:1–14
101. Yan JQ, Wang J, Tissue D, Holaday AS, Allen R, Zhang H (2003) Photosynthesis and seed production under water-deficit conditions in transgenic tobacco plants that overexpress an Arabidopsis ascorbate peroxidase gene. Crop Sci 43:1477–1483
102. Kwon SY, Jeong YJ, Lee HS, Kim JS, Cho KY, Allen RD, Kwak SS (2002) Enhanced tolerances of transgenic tobacco plants expressing both superoxide dismutase and ascorbate peroxidase in chloroplasts against methyl viologen-mediated oxidative stress. Plant Cell Environ 25:873–882
103. Lee SH, Ahsan N, Lee KW, Kim DH, Lee DG, Kwak SS, Kwon SY, Kim TH, Lee BH (2007) Simultaneous overexpression of both Cu Zn superoxide dismutase and ascorbate peroxidase in transgenic tall fescue plants confers increased tolerance to a wide range of abiotic stresses. J Plant Physiol 164:1626–1638
104. Murgia I, Tarantino D, Vannini C, Bracale M, Carravieri S, Soave C (2004) *Arabidopsis thaliana* plants overexpressing thylakoidal ascorbate peroxidase show increased resistance to Paraquat-induced photooxidative stress and to nitric oxide-induced cell death. Plant J 38:940–953
105. Saji H, Aono M, Kubo A, Tanaka K, Kondo N (1997) Paraquat sensitivity of transgenic *Nicotiana tabacum* plants that overproduce a cytosolic ascorbate peroxidase. Phyton-Annales Rei Botanicae 37:259–264
106. Sengupta A, Webb RP, Holaday AS, Allen RD (1993) Overexpression of superoxide dismutase protects plants from oxidative stress—induction of ascorbate peroxidase in superoxide dismutase-overexpressing plants. Plant Physiol 103:1067–1073
107. Rizhsky L, Hallak-Herr E, Van Breusegem F, Rachmilevitch S, Barr JE, Rodermel S, Inze D, Mittler R (2002) Double antisense plants lacking ascorbate peroxidase and catalase are less sensitive to oxidative stress than single antisense plants lacking ascorbate peroxidase or catalase. Plant J 32:329–342
108. Ishikawa T, Sakai K, Takeda T, Shigeoka S (1995) Cloning and expression of cDNA encoding a new type of ascorbate peroxidase from spinach. FEBS Lett 367:28–32
109. Hossain MA, Asada K (1985) Monodehydroascorbate reductase from cucumber is a flavin adenine dinucleotide enzyme. J Biol Chem 260:12920–12926

110. Shimaoka T, Yokota A, Miyake C (2000) Purification and characterization of chloroplast dehydroascorbate reductase from spinach leaves. Plant Cell Physiol 41:1110–1118

111. Elia MR, Borraccino G, Dipierro S (1992) Soluble ascorbate peroxidase from potato tubers. Plant Sci 85:17–21

112. Shimaoka T, Miyake C, Yokota A (2003) Mechanism of the reaction catalyzed by dehydroascorbate reductase from spinach chloroplasts. Eur J Biochem 270:921–928

113. Urano J, Nakagawa T, Maki Y, Masumura T, Tanaka K, Murata N, Ushimaru T (2000) Molecular cloning and characterization of a rice dehydroascorbate reductase. FEBS Lett 466:107–111

114. Chen Z, Young TE, Ling J, Chang SC, Gallie DR (2003) Increasing vitamin C content of plants through enhanced ascorbate recycling. Proc Nat Acad Sci USA 100:3525–3530

115. Qin AG, Shi QH, Yu XC (2011) Ascorbic acid contents in transgenic potato plants overexpressing two dehydroascorbate reductase genes. Mol Biol Rep 38:1557–1566

116. Eltelib HA, Fujikawa Y, Esaka M (2012) Overexpression of the acerola (*Malpighia glabra*) monodehydroascorbate reductase gene in transgenic tobacco plants results in increased ascorbate levels and enhanced tolerance to salt stress. S Afr J Bot 78:295–301

117. Chen Z, Gallie DR (2004) The ascorbic acid redox state controls guard cell signaling and stomatal movement. Plant Cell 16:1143–1162

118. Chen Z, Gallie DR (2005) Increasing tolerance to ozone by elevating foliar ascorbic acid confers greater protection against ozone than increasing avoidance. Plant Physiol 138:1673–1689

119. Grantz A, Brummell DA, Bennett AB (1995) Ascorbate free radical reductase mRNA levels are induced by wounding. Plant Physiol 108:411–418

120. Yoon HS, Lee H, Lee IA, Kim KY, Jo JK (2004) Molecular cloning of the monodehydroascorbate reductase gene from *Brassica campestris* and analysis of its mRNA level in response to oxidative stress. Biochimica et Biophysica Acta-Bioenergetics 1658:181–186

121. Deleonardis S, Delorenzo G, Borraccino G, Dipierro S (1995) A specific ascorbate free radical reductase isozyme participates in the regeneration of ascorbate for scavenging toxic oxygen species in potato tuber mitochondria. Plant Physiol 109:847–851

122. Leterrier M, Corpas FJ, Barroso JB, Sandalio LM, del Rio LA (2005) Peroxisomal monodehydroascorbate reductase. Genomic clone characterization and functional analysis under environmental stress conditions. Plant Physiol 138:2111–2123

123. Sano S, Tao S, Endo Y, Inaba T, Hossain A, Hossain MA, Miyake C, Matsuo M, Aoki H, Asada K, Saito K (2005) Purification and cDNA cloning of chloroplastic monodehydroascorbate reductase from spinach. Biosci Biotechnol Biochem 69:762–772

124. Sano S, Asada K (1994) cDNA cloning of monodehydroascorbate radical reductase from cucumber: a high-degree of homology in terms of amino acid sequence between this enzyme and bacterial flavoenzymes. Plant Cell Physiol 35:425–437

125. Murthy SS, Zilinskas BA (1994) Molecular cloning and characterization of a cDNA encoding pea monodehydroascorbate reductase. J Biol Chem 269:31129–31133

126. Arrigoni O (1994) Ascorbate system in plant development. J Bioenerg Biomembr 26:407–419

127. Loewus FA, Loewus MW (1987) Biosynthesis and metabolism of ascorbic acid in plants. Crit Rev Plant Sci 5:101–119

128. Lorence A, Chevone BI, Mendes P, Nessler CL (2004) myo-inositol oxygenase offers a possible entry point into plant ascorbate biosynthesis. Plant Physiol 134:1200–1205

129. Kanter U, Usadel B, Guerineau F, Li Y, Pauly M, Tenhaken R (2005) The inositol oxygenase gene family of Arabidopsis is involved in the biosynthesis of nucleotide sugar precursors for cell-wall matrix polysaccharides. Planta 221:243–254

130. Endres S, Tenhaken R (2009) Myoinositol oxygenase controls the level of myoinositol in Arabidopsis, but does not increase ascorbic acid. Plant Physiol 149:1042–1049

131. Jain AK, Nessler CL (2000) Metabolic engineering of an alternative pathway for ascorbic acid biosynthesis in plants. Mol Breed 6:73–78

132. Radzio JA, Lorence A, Chevone BI, Nessler CL (2003) L-Gulono-1,4-lactone oxidase expression rescues vitamin C-deficient Arabidopsis (*vtc*) mutants. Plant Mol Biol 53:837–844

133. Hemavathi C, Upadhyaya P, Akula N, Young KE, Chun SC, Kim DH, Park SW (2010) Enhanced ascorbic acid accumulation in transgenic potato confers tolerance to various abiotic stresses. Biotechnol Lett 32:321–330

134. Agius F, Gonzalez-Lamothe R, Caballero JL, Munoz-Blanco J, Botella MA, Valpuesta V (2003) Engineering increased vitamin C levels in plants by overexpression of a D-galacturonic acid reductase. Nat Biotechnol 21:177–181

135. Karpinski S, Escobar C, Karpinska B, Creissen G, Mullineaux PM (1997) Photosynthetic electron transport regulates the expression of cytosolic ascorbate peroxidase genes in Arabidopsis during excess light stress. Plant Cell 9:627–640

136. Caldwell CR, Turano FJ, McMahon MB (1998) Identification of two cytosolic ascorbate peroxidase cDNAs from soybean leaves and characterization of their products by functional expression in *E.coli*. Planta 204:120–126

137. Morimura Y, Iwamoto K, Ohya T, Igarashi T, Nakamura Y, Kubo A, Tanaka K, Ikawa T (1999) Light-enhanced induction of ascorbate peroxidase in Japanese radish roots during postgerminative growth. Plant Sci 142:123–132

138. Lee SK, Kader AA (2000) Preharvest and postharvest factors influencing vitamin C content of horticultural crops. Postharvest Biol Technol 20:207–220

Chapter 8
Regulation of Ascorbate Synthesis in Plants

Different biosynthesis pathways as well as oxidization, recycling of ascorbate jointly constitute the complex network of ascorbate metabolism. The huge effort has been made on identifying the structural enzymes and the delineation of the catalytic pathways. The critical enzymes and their encoding genes were successively purified or isolated from plant species. However, the regulation of the specific pathway and the interaction between different processes remains largely unknown.

Regulators or transcription factors are reportedly involved in ascorbate biosynthesis and metabolism. In the meantime, ascorbate biosynthesis and metabolism are modulated by plant growth development and environmental conditions [1]. The ascorbate accumulated to different levels in different plant development stages. The external factors like light can influence the ascorbate accumulation significantly. Elucidating the regulation mechanism of the ascorbate biosynthesis and metabolism will facilitate the metabolic engineering for improving the important antioxidant.

8.1 Growth and Postharvest Condition

The contents of ascorbate accumulation can vary in different plant species and different tissues of the same plant species. The ascorbate content also varies with developmental stages. Komatsuna plants harvested when the length of the longest leaf was about 25 cm contained the highest concentration of ascorbate [2].

The ascorbate and its biosynthesis and metabolism related enzymes can help plant to response positively to environmental stress. The growth condition and environmental factors will in turn influence the ascorbate level and activity of its metabolism related enzymes.

Plants are frequently confronted with abiotic stress throughout their life cycle, which can affect the ascorbate dynamic in plants. Under salt stress (10–50 mM/kg NaCl in soil), ascorbate content in plants increased with NaCl concentration in

Y. Zhang, *Ascorbic Acid in Plants*, SpringerBriefs in Plant Science,
DOI: 10.1007/978-1-4614-4127-4_8, © The Author 2013

soil, with a significant correlation with the chloride content in plant [3]. In sorghum, drought decreased cytosolic activities of APX, MDHAR, and increased cytosolic DHAR activity. In sunflower, drought increased chloroplastic APX activity and cytosolic activity of DHAR [4]. The enzyme involved in ascorbate metabolism is found to be regulated by oxygen. In *Cucurbita pepo*, the AO activity was elicited in dark by increased oxygen concentration, and decreased during germination under hypoxic conditions. This result suggests that AO activity could be part of a dynamic system for oxygen management in plants [5].

The relatively lower concentrations of nutrient solutions can help raise the ascorbate level in hydroponically cultured plants. The ascorbate accumulation in plants increased steadily as the nitrogen levels decreased from 270 kg N/ha to 0 kg N/ha [2]. Leaves of nitrogen-sufficient plants often had as much or more total ascorbate than leaves of nitrogen-limited plants. However, the leaves of nitrogen-limited plants displayed considerably higher ratios of ascorbate to dehydroascorbate in the light periods than did leaves of nitrogen-sufficient plants [6]. Thus, in order to prevent negative effects on the ascorbate content in plant product, it is advisable to avoid excessive application of nitrogen fertilizer when cultivating crops [7]. This is consistent with the fact that the ascorbate content in organically cultivated vegetables is higher than that in conventionally produced vegetables with chemical fertilizer [8].

APX enzyme, which is responsible for ascorbate metabolism as well as ROS scavenging, is shown to be regulated by various environmental clues. The micronutrient deficiencies can alter the APX activities depending on the severity and the type of the deficiency stress. The APX activities declined drastically in plants with deficiency of Mn or Zn, but remained significantly higher in plants grown without Cu than those of control plants [9]. The APX activity is affected by selenium content in plants. With the increasing plant selenium concentration, the APX activity was decreased in wheat but increased in oilseed rape [10]. APX activity in plants is enhanced in salt stress, thought its mRNA level is relatively stable [11]. The heavy metals can cause alteration in ascorbate content as well as the APX enzyme in plants. Ascorbate level and APX activity exhibited an initial increase and subsequent decrease in response to heavy metal stress, resulting in significantly lowered final ascorbate content under high metal concentration [12].

The herbicide showed a dual effect on ascorbate accumulation in plants. Low herbicide (atrazine, trifluralin, and metolachlor) levels (less than 0.1 μg/mL) led to increased total ascorbate suggesting a stimulatory effect on ascorbate synthesis occurred, while at higher herbicide concentrations (greater than or equal to 0.1 μg/mL) a notable decline in total ascorbate and increase in the oxidized form, dehydroascorbate occurred [13]. Treatment of another herbicide paraquat also resulted in rapid decrease of ascorbate content in plant leaves [14].

The ascorbate metabolism in plants is also shown to be affected by pathogenic bacteria infection. Exopolysaccharides produced by plant pathogenic bacteria induce an alteration in the metabolism of ascorbate as well as the increase in hydrogen peroxide production in plant cells [15]. This is reminiscent of the fact that many defense genes, particularly those that encode pathogenesis-related proteins, are activated in the ascorbate defective mutant *vtc1* [16].

Herbivores will also probably influence both de novo synthesis and redox cycling of ascorbate in host plants, thereby potentially altering the nutritional value of crops and their susceptibility to pests. The recent development of genetically modified crops with enhanced ascorbate content provides both an impetus and a tool set for further studies on the role of ascorbate in plant–insect interactions [17].

Ascorbate is not stable and easily oxidized, so the processing and storage of plant product will inevitably cause looses in ascorbate. The ascorbate concentration in the subtropical plant species, *Cyphostemma digitatum*, 49.50 ± 0.01 mg/100 g in the raw material and 20.30 ± 0.02 mg/100 g in the processed material, was affected negatively by processing, retaining only 41 % after processing [18]. Boiling of aqueous mixtures of vegetables reduced the ascorbate content by 20–43 % [19]. Also, the content of ascorbate as well as other nutrients decreased in plant products (black currant nectars) during the storage despite of the potential protection of plant extract additives [20].

8.2 Light Regulation

Light is shown to play crucial role in regulating the ascorbate biosynthesis and metabolism of high plants. The ascorbate content in leaves under high intensity of illumination is usually higher than that in shaded leaves. L-galactono-1,4-lacton feeding resulted in ascorbate increase in Arabidopsis leaves under light, while in dark conditions ascorbate content in leaves remained unchanged when fed with L-galactono-1,4-lacton. This association of light with ascorbate synthesis was reported in Arabidopsis [21–23], radish [24], rice [25], tobacco [23], apple [26, 27] and kiwifruit [28]. A (CO_2)-C^{13} feeding assay showed that the ascorbate in leaves is synthesized from newly fixed C and partly from the reserved C, indicating that light-dependent photosynthesis is a pivotal process for ascorbate biosynthesis [29]. The regulating mechanism of light upon ascorbate metabolism remains unclear so far.

The ascorbate in plants shows daily oscillations, providing a possible light regulation rhythm in ascorbate production. This daily fluctuations may also be affected by plant development stages or other environmental factors [30]. Reducing fruit irradiance strongly decreased the reduced ascorbate content. Leaf shading delayed fruit ripening and promoted the accumulation of oxidized ascorbate in green fruit, whereas it decreased the reduced ascorbate content in orange fruits, suggesting that complex regulation of ascorbate metabolism depends on irradiance of fruit and leaf as well as fruit ripening stage [31].

The subcellular distribution of ascorbate is also modulated by high light as revealed by immunocytochemical determination. High light intensities resulted in a strong increase in overall ascorbate labeling density. Interestingly, the strongest compartment-specific increase was found in vacuoles (4 folds) and in plastids (2 folds). Ascorbate-specific labeling was restricted to the matrix of mitochondria and to the stroma of chloroplasts in control plants but was also detected in the lumen of thylakoids after high light exposure [23].

Light helps to regulate the expression of genes involved in ascorbate metabolism in plants. High intensity of light improved the expression abundance of *GMP* and *GLDH* in tobacco cells [32]. In Arabidopsis, several genes involved in ascorbate biosynthesis were shown to be regulated by light, e.g. *VTC2* [33, 34], *PMI* [35], *GalDH* [22] and *GLDH* [36]. In cucurbit, abundant mRNA, protein and enzyme activity of AO were observed in illuminated leaves. AO activity was found to be proportional to light intensity. The light effect was rapidly reversed in dark and activity remained low throughout the dark period [5]. Gene expression of *GalDH* and *GLDH* was also regulated by light in apple. Semishading of apple tree resulted in significant decrease in the expression level of both *GalDH* and *GLDH* [26]. Morimura et al. [24] found that during the early growth stage of Japanese radish after germination, the *cAPX* gene expression in roots was rapidly induced possibly by light [24]. Light intensity regulated the expression of *GPP* and *GLDH* in rice, and light responsive cis-elements, GT1 box and TGACG motif, were found in the promoter region of *GPP* and *GLDH*, suggesting ascorbate synthesis in plants is regulated by light at the transcriptional level [25].

Photosynthesis was found to be tightly connected with the ascorbate biosynthesis. Recent study indicates that photosynthetic electron transfer (PET) is one of the most important influencing factors for ascorbate synthesis. Upon treatment of diuron (DCMU) and atrazine (ATZ), the inhibitors of photosynthetic electron transfer chain, the ascorbate in Arabidopsis leaves decreased significantly and the down-regulated expression was observed in genes responsible for ascorbate biosynthesis pathway, e.g. *GMP*, *GPP*, *VTC2* and *GLDH* [37]. Further evidence showed that redox status of photosynthetic electron transport chain contributed to the conversion from L-galactono-1,4-lactone to ascorbate [38].

The light is shown to control ascorbate synthesis and accumulation in plants via interaction with respiration. The hypothesis that respiration act in regulating the ascorbate synthesis was put forward for the first time in 2003 based on three observations [39]. Firstly, GLDH, the last step enzyme for ascorbate biosynthesis, was localized in mitochondrial complex I, which is regulated by electron transfer. Secondly, during ascorbate formation, GLDH needs the oxidized cytochrome c as the electron acceptor. And thirdly, both respiration and ascorbate synthesis respond to hormone treatment coordinately. This hypothesis is supported by more evidence. The ascorbate content in Arabidopsis leaves increase with the respiration under high light, and overexpressing gene encoding mitochondrial alternative oxidase (AOX) resulted in enhanced accumulation of ascorbate in Arabidopsis [21]. Thus the interaction of light and respiration could be crucial determinant for the plant capacity to produce and accumulate ascorbate.

8.3 Plant Growth Regulator

The ascorbate level in plants could be modulated by exogenous plant growth regulators. A significant enhancement in non-enzymatic antioxidant contents was observed in all sampling days in *A. paniculata* plants under ABA and GA treatments. Ascorbate content was increased significantly under the growth regulator treatments in leaves, stem and roots of *A. paniculata*. The activities of APX were increased by ABA and GA treatments in the leaves, stem and roots [40].

Exogenous ascorbate, in turn, can modify the physiological effects of ABA on plants. ABA modulates the hydraulic properties of roots by increasing root water flux, and ascorbate can partially or, at 100 μmol/L, completely inhibit the ABA stimulation of water flux [41]. This is reminiscent that ascorbate modulates the plant growth and development in concert with phytohormones at the transcriptional level [16].

APX activity can be inhibited by salicylic acid (SA) and 2,6-dichloroisonicotinic acid, two inducers of plant defense responses [42]. The inhibition of APX by SA supports the hypothesis that SA-induced defense responses are mediated, in part, through elevated hydrogen peroxide levels or coupled perturbations of the cellular redox state. The APX activity was also shown to be inhibited by benzothiadiazole, a functional analogue of SA [43]. Catechin treatment also affects APX activity in close coordination with SOD and CAT. Exogenous catechin can markedly reduce the waterlogging injury in leaves and roots of tomato by enhancing free radical scavenging system sufficiently to lower hydrogen peroxide and superoxide concentrations [44].

The ethephon treatment produced a rapid decrease in ascorbate content and an increase in its oxidised state in spinach. Both ascorbate synthesis and recovery from oxidised forms were affected by ethylene. In addition, leaves from *Arabidopsis* ethylene insensitive mutants (*ein2-1*, *ein3-1* and *ein4*) exhibited higher ascorbate content than leaves from wild-type plants. The Arabidopsis mutant with constitutive triple response (*ctr1-1*) showed lower leaf ascorbate content than wild-type plants. The APX enzyme, which is responsible for ascorbate oxidization and metabolism, is found to be stimulated by ethylene treatment [45]. These results demonstrate that ethylene is an important factor controlling ascorbate content in plants [46].

It is interesting that SA-deficient plants adapt to RNA virus infections better, which show a lighter symptom and have less ROS accumulation. This alleviated symptom is supposed to be attributed to the higher level ascorbate in SA-deficient plants [47].

It should be noted that the ascorbate is the coenzyme for the ACC oxidase and GA2-oxidase, and thus involved in the biosynthesis of ethylene and GA. This is consistent with the regulatory role of plant growth regulator on ascorbate and the function of ascorbate in plant growth and development.

8.4 Jasmonates Regulation

Jasmonic acid (JA) and methyl jasmonate (MeJA), generally termed as jasmonates, are important signals in response to abiotic stress in plants. Various stress conditions lead to JA accumulation. JA and MeJA help plants to respond to biotic and abiotic stress through activating series of gene expression in plants. Emerging evidences indicate that jasmonates participate in regulating ascorbate biosynthesis in plants.

Jasmonates helps to defend plants against environmental stress through activating antioxidant-related pathway, such as ascorbate metabolism. Several genes related to ascorbate biosynthesis and recycling, e.g. *GMP* and *GGP* for the synthesis and *MDHAR* and *DHAR* for the recycling were up-regulated by treatment of JA or MeJA in Arabidopsis, as revealed by cDNA microarray [48]. In addition, MeJA treatment resulted in the enhancement of ascorbate content as well as increase in activity of DHAR and APX. The regulation effect of MeJA on ascorbate synthesis was supported by emerging evidence. The radioactive labeled ^{14}C-mannose combined with HPLC and transcriptional profile analysis showed that the ascorbate content in cell suspensions of Arabidopsis and tobacco treated with MeJA was higher that that of control [49]. The expression abundance of the *GME* and one putative gene encoding L-gulonolactone dehydrogenase/oxidase in tobacco cell suspension treated with MeJA was higher than that of untreated control. Thus the MeJA might promote ascorbate synthesis by up-regulating ascorbate biosynthesis related genes.

The JA level, the transcriptional abundance as well as enzymatic activities of APX, GR, MDHAR, DHAR, GalLDH, and the contents of ascorbate and glutathione were increased by water stress. Water stress-induced JA is a signal that leads to the regulation of ascorbate and glutathione metabolism and has important role for acquisition of water stress tolerance in plants [50]. In tomato and Arabidopsis, certain mutations that impair JA metabolism and signaling influence foliar ascorbate levels, suggesting that endogenous JAs may regulate steady-state ascorbate. JA effects on ascorbate accumulation vary with plant species. JA enhances ascorbate accumulation in Arabidopsis, but suppresses ascorbate levels in tomato, demonstrating the complexity of ascorbate regulation [51].

8.5 Feedback Regulation

Several evidences support the hypothesis of feedback regulation in the process of ascorbate synthesis. The synthesis rate from [U-^{14}C] D-glucose to ascorbate decreased dramatically with the increase of ascorbate pool size in pea seedlings [52]. The feedback suppression on the ascorbate synthesis was also observed in Arabidopsis cell suspensions. The amount of ascorbate synthesized from [^{14}C]D-mannose decreased in the presence of ascorbate [53]. The enzyme activity of L-galactose dehydrogenase was feedback suppressed by the ascorbate in spinach [54].

Ascorbate of 1 mmol/L concentration results in 41 % decreasing in GalDH enzyme activity [54], indicating the feedback regulation mechanism in the ascorbate bio-synthesis. The exogenous loading of ascorbate in tobacco cell suspension resulted in significantly decreased expression level of both *GMP* and *GLDH* [32]. All this results demonstrate that ascorbate synthesis in plant cells is subject to feedback suppression at various steps, though the mechanism remains unclear so far. And this feedback regulation may help to keep homeostasis of ascorbate synthesis in plants.

8.6 Coordination and Compensation

When we use gene overexpression or suppression to modulate the biosynthesis of ascorbate in plants, we found the phenomenon of coordination and compensation. Transgenic tomato overexpressing the D-galacturonate reductase gene from strawberry (*FaGalUR*) showed increased ascorbate levels in ripe fruits compared to wild-type control. Interestingly, expression level of *GLDH*, gene encoding the last step enzyme for ascorbate synthesis, in the transgenic tomato was also improved in correlation to the ascorbate level (unpublished data). This may be interpreted as the coordination regulation for ascorbate. When plants are subjected to beneficial modulation (e.g. increasing ascorbate synthesis and accumulation), the genes of the related biological process are activated to reinforce the metabolic advantages to plants.

This coordination regulation for ascorbate synthesis could also be observed in transgenic tomato overexpressing the *GME* gene, in which decreased expression of *AO* gene as well as the increased expression of *GME* gene was associated with the improved ascorbate accumulation (unpublished data). At same time, in the transgenic tomato overexpressing *GMP* gene, the enhanced expression of *GMP* was accompanied by the elevated expression level of genes in the D-Man/L-Gal pathway (Fig. 8.1, denoted in open box).

Surprisingly, the suppressed expression of *GMP* by RNA interference in tomato resulted in the elevated expression of genes in D-Man/L-Gal pathway (Fig. 8.2, denoted in open box). This could be interpreted as the compensation machinery for ascorbate synthesis and accumulation to keep the ascorbate homeostasis in plants. When plants are confronted with disadvantages (e.g. decreasing ascorbate synthesis and accumulation), the alternative pathways or genes may start working to counteract the adverse effect and cope with the challenges.

Antisense suppression of *GalDH* enzyme in plants did not result in alteration of ascorbate level in Arabidopsis under weak light, and only slight decrease in ascorbate under strong light was observed. This slight decreasing of ascorbate could be neglected when compared to the dramatic decreasing in GalDH enzyme activity [22]. The *GalDH* suppressed plants may activate compensation machinery to make up the decreasing GalDH activity and keep the ascorbate homeostasis in plants.

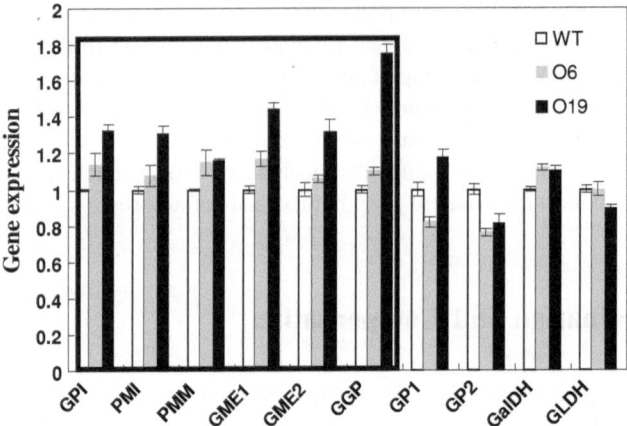

Fig. 8.1 *GMP* overexpression in tomato activates the expression the genes involved in D-Man/L-Gal pathway of ascorbate synthesis (coordination regulation) WT. wild-type; O6 and O19 denote two overexpression lines with *GMP* gene (courtesy of Professor Zhibiao Ye and Dr. Chanjuan Zhang)

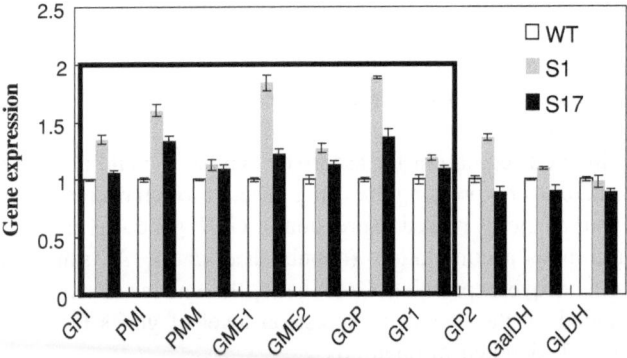

Fig. 8.2 Suppressed *GMP* expression in tomato is accompanied by elevated expression of genes involved in D-Man/L-Gal pathway of ascorbate synthesis (compensation regulation) WT. wild-type; S1 and S17 denote two transgenic lines line with *GMP* suppressed by RNA interference (courtesy of Professor Zhibiao Ye and Dr. Chanjuan Zhang)

This compensation mechanism is also true in terms of APX activity regulation. APX is a well-known enzymatic antioxidant in addition to its crucial role in ascorbate metabolism. The rice mutants double silenced for cytosolic APXs (APX1/2) activate other peroxidases (catalase and guaiacol peroxidase) enabling the mutants to cope with abiotic stress, with the similar stress response capacity of wild-type plants [55].

Compensation mechanism also applies in transgenic plants with double antisense suppression of APX and CAT, which are less sensitive to oxidative stress than single antisense plants lacking APX or CAT [56]. The double antisense plants lacking the two major hydrogen peroxide-detoxifying enzymes, APX and CAT, activate an alternative/redundant defense mechanism that compensates for the lack of APX and CAT. A similar mechanism was not activated in single antisense plants that lacked APX or CAT, paradoxically rendering these plants more sensitive to oxidative stress compared to double antisense plants. The coordinated induction of metabolic and defense genes, coupled with the suppression of photosynthetic activity, may compensate for the lack of APX.

The plant genome is shown to be highly redundant to keep its biological process flexible in response to various environments. The ascorbate synthesis is supposed to be regulated by compensation machinery under certain circumstances.

8.7   Transcription Factor

The Ascorbic acid Mannose pathway Regulator 1 (AMR1), the first regulator for ascorbate biosynthesis, was isolated from ozone sensitive Arabidopsis mutants generated by activation tagging [57]. *AMR1* encodes a F-box protein with a F-box motif on its N terminal. AMR1 shows high identity with other F-box proteins in human and Arabidopsis. On the C terminal of AMR1 there is a DUF295 domain without known function composed of 48 amino acids (MMEVKSLGDKAF-VIATDTCFSVLAHEFYGCLENAIYFTDDT). Several functionally unknown proteins have been shown to contain both F-box and DUF295 in Arabidopsis, indicating more than one AMR1-like protein in Arabidopsis. The *AMR1* mutation generated by activation tagging in Arabidopsis resulted in 40 % of ascorbate content of the wild-type control, while the T-DNA insertional mutant in Arabidopsis led to 2–3 folds of the ascorbate content of wild-type. The expression of most genes involved in D-Man/L-Gal pathway, including *GMP*, *GGP*, *GPP*, *GME*, *GalDH* and *GLDH*, was down-regulated in the ozone sensitive mutant with activation tagging, but up-regulated in the T-DNA insertional mutant. It is proposed that AMR1 affects ascorbate biosynthesis by negatively regulating the genes involved in D-Man/L-Gal pathway. The *MIOX4* gene, involved in the myoinositol pathway, however, was not regulated by AMR1. The expression of *AMR1* gene was regulated by plant growth and development as well as the environmental factors. The expression abundance of *AMR1* increased with the leaf development, and decreased with the light intensity.

The F-box domain in AMR1 shows high homology with its counterparts in other plant species, e.g. ORE9 and UFO. F-box protein is a part of the SCF-ubiquitin-E3 ligase complex that is involved in recognition of both the E2 protein, containing activated ubiquitin, and the substrate targeted for ubiquitination. Selective protein degradation by the ubiquitin-proteosome pathway is a powerful regulatory mechanism in a wide variety of cellular processes [57]. AMR1, as a part of SCF ubiquitin ligase complex, could possibly interact with a positive regulator, and regulate its protein amount and stability, and thus modulates the ascorbate biosynthesis.

AMR1 is a regulator that can simultaneously modulate the expression of several genes in the D-Man/L-Gal pathway. However, other regulatory genes may exist, as ascorbate synthesis is subject to delicate regulation by environment as well as plant growth and development.

References

1. Smirnoff N, Wheeler GL (2000) Ascorbic acid in plants: biosynthesis and function. Crit Rev Plant Sci 19:267–290
2. Miyajima D (1994) Effects of concentration of nutrient solution, plant size at harvest, and light condition before harvest on the ascorbic acid and sugar concentrations in leaves of hydroponically grown komatsuna (*Brassica campestris* L. rapifera group). J Jpn Soc Hortic Sci 63:567–574
3. Telesinski A, Nowak J, Smolik B, Dubowska A, Skrzypiec N (2008) Effect of soil salinity on activity of antioxidant enzymes and content of ascorbic acid and phenols in bean (*Phaseolus vulgaris* L.) Plants. J Elementol 13:401–409
4. Zhang JX, Kirkham MB (1996) Enzymatic responses of the ascorbate-glutathione cycle to drought in sorghum and sunflower plants. Plant Sci 113:139–147
5. De Tullio MC, Cirad S, Liso R, Arrigoni O (2007) Ascorbic acid oxidase is dynamically regulated by light and oxygen. A tool for oxygen management in plants? J Plant Physiol 164:39–46
6. Robinson JM (1997) The influence of elevated foliar carbohydrate levels on the ascorbate: dehydroascorbate redox ratios in nitrogen-limited spinach and soybean plants. Int J Plant Sci 158:442–450
7. Ibrahim MH, Jaafar HZE, Rahmat A, Rahman ZA (2012) Involvement of nitrogen on flavonoids, glutathione, anthocyanin, ascorbic acid and antioxidant activities of malaysian medicinal plant *Labisia pumila* Blume (Kacip Fatimah). Int J Mol Sci 13:393–408
8. Citak S, Sonmez S (2011) Monitoring of plant growth, yield, vitamin C and nitrate concentrations in organically and conventionally grown white head cabbage (*Brassica oleracea* var. capitata) in two successive seasons. J Agric Food Chem 9:501–505
9. Yu Q, Osborne L, Rengel Z (1998) Micronutrient deficiency changes activities of superoxide dismutase and ascorbate peroxidase in tobacco plants. J Plant Nutr 21:1427–1437
10. Kakewski K, Nowak J, Ligocki M (2008) Effects of selenium content in green parts of plants on the amount of ATP and ascorbate-glutathione cycle enzyme activity at various growth stages of wheat and oilseed rape. J Plant Physiol 165:1011–1022
11. Lopez F, Vansuyt G, CasseDelbart F, Fourcroy P (1996) Ascorbate peroxidase activity, not the mRNA level, is enhanced in salt-stressed *Raphanus sativus* plants. Physiol Plant 97:13–20
12. Huang GY, Wang YS, Sun CC, Dong JD, Sun ZX (2010) The effect of multiple heavy metals on ascorbate, glutathione and related enzymes in two mangrove plant seedlings (*Kandelia candel* and *Bruguiera gymnorrhiza*). Oceanol Hydrobiol Stud 39:11–25
13. Lytle TF, Lytle JS (1997) Ascorbate: a biomarker of herbicide stress in wetland plants. ACS Symp Ser 664:106–113
14. Saji H, Aono M, Kubo A, Tanaka K, Kondo N (1997) Paraquat sensitivity of transgenic *Nicotiana tabacum* plants that overproduce a cytosolic ascorbate peroxidase. Phyton-Annales Rei Botanicae 37:259–264
15. de Pinto MC, Lavermicocca P, Evidente A, Corsaro MM, Lazzaroni S, De Gara L (2003) Exopolysaccharides produced by plant pathogenic bacteria affect ascorbate metabolism in *Nicotiana tabacum*. Plant Cell Physiol 44:803–810
16. Pastori GM, Kiddle G, Antoniw J, Bernard S, Veljovic-Jovanovic S, Verrier PJ, Noctor G, Foyer CH (2003) Leaf vitamin C contents modulate plant defense transcripts and regulate genes that control development through hormone signaling. Plant Cell 15:939–951

17. Goggin FL, Avila CA, Lorence A (2010) Vitamin C content in plants is modified by insects and influences susceptibility to herbivory. BioEssays 32:777–790
18. Al-Duais M, Hohbein J, Werner S, Bohm V, Jetschke G (2009) Contents of vitamin C, carotenoids, tocopherols, and tocotrienols in the subtropical plant species *Cyphostemma digitatum* as affected by processing. J Agric Food Chem 57:5420–5427
19. Ogunlesi M, Okiei W, Azeez L, Obakachi V, Osunsanmi M, Nkenchor G (2010) Vitamin C contents of tropical vegetables and foods determined by voltammetric and titrimetric methods and their relevance to the medicinal uses of the plants. Int J Electrochem Sci 5:105–115
20. Kalisz S, Scibisz I (2010) Effect of plant extract additives on the content of total polyphenols, anthocyanins, L-ascorbic acid, and antioxidant capacity of black currant nectars. Zywnosc-Nauka Technologia Jakosc 17:45–55
21. Bartoli CG, Yu JP, Gomez F, Fernandez L, McIntosh L, Foyer CH (2006) Inter-relationships between light and respiration in the control of ascorbic acid synthesis and accumulation in *Arabidopsis thaliana* leaves. J Exp Bot 57:1621–1631
22. Gatzek S, Wheeler GL, Smirnoff N (2002) Antisense suppression of L-galactose dehydrogenase in *Arabidopsis thaliana* provides evidence for its role in ascorbate synthesis and reveals light modulated L-galactose synthesis. Plant J 30:541–553
23. Zechmann B, Stumpe M, Mauch F (2011) Immunocytochemical determination of the subcellular distribution of ascorbate in plants. Planta 233:1–12
24. Morimura Y, Iwamoto K, Ohya T, Igarashi T, Nakamura Y, Kubo A, Tanaka K, Ikawa T (1999) Light-enhanced induction of ascorbate peroxidase in Japanese radish roots during postgerminative growth. Plant Sci 142:123–132
25. Fukunaga K, Fujikawa Y, Esaka M (2010) Light regulation of ascorbic acid biosynthesis in rice via light responsive cis-elements in genes encoding ascorbic acid biosynthetic enzymes. Biosci Biotechnol Biochem 74:888–891
26. Li MJ, Ma FW, Shang PF, Zhang M, Hou CM, Liang D (2009) Influence of light on ascorbate formation and metabolism in apple fruits. Planta 230:39–51
27. Li Q, Li BH, Kronzucker HJ, Shi WM (2010a) Root growth inhibition by NH_4^+ in Arabidopsis is mediated by the root tip and is linked to NH_4^+ efflux and GMPase activity. Plant Cell Environ 33:1529–1542
28. Li MJ, Ma FW, Liu J, Li JA (2010b) Shading the whole vines during young fruit development decreases ascorbate accumulation in kiwi. Physiol Plant 140:225–237
29. Yoneyama T, Yasuda M, Sato S, Takebe M (1997) $^{13}CO_2$ feeding studies on the metabolism of carbohydrates, ascorbate and oxalate in spinach (*Spinacia oleracea* L.). Soil Sci Plant Nutr 43:1147–1151
30. Lee EH (1991) Plant resistance mechanisms to air pollutants: rhythms in ascorbic acid production during growth under ozone stress. Chronobiol Int 8:93–102
31. Gautier H, Massot C, Stevens R, Serino S, Genard M (2009) Regulation of tomato fruit ascorbate content is more highly dependent on fruit irradiance than leaf irradiance. Ann Bot 103:495–504
32. Tabata K, Takaoka T, Esaka M (2002) Gene expression of ascorbic acid-related enzymes in tobacco. Phytochemistry 61:631–635
33. Dowdle J, Ishikawa T, Gatzek S, Rolinski S, Smirnoff N (2007) Two genes in *Arabidopsis thaliana* encoding GDP-L-galactose phosphorylase are required for ascorbate biosynthesis and seedling viability. Plant J 52:673–689
34. Muller-Moule P (2008) An expression analysis of the ascorbate biosynthesis enzyme VTC2. Plant Mol Biol 68:31–41
35. Maruta T, Yonemitsu M, Yabuta Y, Tamoi M, Ishikawa T, Shigeoka S (2008) Arabidopsis phosphomannose isomerase 1, but not phosphomannose isomerase 2, is essential for ascorbic acid biosynthesis. J Biol Chem 283:28842–28851
36. Tamaoki M, Mukai F, Asai N, Nakajima N, Kubo A, Aono M, Saji H (2003) Light-controlled expression of a gene encoding L-galactono-gamma-lactone dehydrogenase which affects ascorbate pool size in *Arabidopsis thaliana*. Plant Sci 164:1111–1117

37. Yabuta Y, Mieda T, Rapolu M, Nakamura A, Motoki T, Maruta T, Yoshimura K, Ishikawa T, Shigeoka S (2007) Light regulation of ascorbate biosynthesis is dependent on the photosynthetic electron transport chain but independent of sugars in Arabidopsis. J Exp Bot 58:2661–2671
38. Yabuta Y, Maruta T, Nakamura A, Mieda T, Yoshimura K, Ishikawa T, Shigeoka S (2008) Conversion of L-galactono-1,4-lactone to L-ascorbate is regulated by the photosynthetic electron transport chain in Arabidopsis. Biosci Biotechnol Biochem 72:2598–2607
39. Chew O, Whelan J, Millar AH (2003) Molecular definition of the ascorbate-glutathione cycle in Arabidopsis mitochondria reveals dual targeting of antioxidant defenses in plants. J Biol Chem 278:46869–46877
40. Anuradha VE, Jaleel CA, Salem MA, Gomathinayagam M, Panneerselvam R (2010) Plant growth regulators induced changes in antioxidant potential and andrographolide content in *Andrographis paniculata* Wall. ex Nees. Pestic Biochem Physiol 98:312–316
41. Aroca R (2006) Exogenous catalase and ascorbate modify the effects of abscisic acid (ABA) on root hydraulic properties in *Phaseolus vulgaris* L. plants. J Plant Growth Regul 25:10–17
42. Durner J, Klessig DF (1995) Inhibition of ascorbate peroxidase by salicylic-acid and 2,6-dichloroisonicotinic acid, 2 inducers of plant defense responses. Proc Nat Acad Sci U S A 92:11312–11316
43. Wendehenne D, Durner J, Chen ZX, Klessig DF (1998) Benzothiadiazole, an inducer of plant defenses, inhibits catalase and ascorbate peroxidase. Phytochemistry 47:651–657
44. Yiu JC, Tseng MJ, Liu CW (2011) Exogenous catechin increases antioxidant enzyme activity and promotes flooding tolerance in tomato (*Solanum lycopersicum* L.). Plant Soil 344:213–225
45. Mehlhorn H (1990) Ethylene-promoted ascorbate peroxidase activity protects plants against hydrogen peroxide, ozone and paraquat. Plant Cell Environ 13:971–976
46. Gergoff G, Chaves A, Bartoli CG (2010) Ethylene regulates ascorbic acid content during dark-induced leaf senescence. Plant Sci 178:207–212
47. Wang SD, Zhu F, Yuan S, Yang H, Xu F, Shang J, Xu MY, Jia SD, Zhang ZW, Wang JH, Xi DH, Lin HH (2011) The roles of ascorbic acid and glutathione in symptom alleviation to SA-deficient plants infected with RNA viruses. Planta 234:171–181
48. Sasaki-Sekimoto Y, Taki N, Obayashi T, Aono M, Matsumoto F, Sakurai N, Suzuki H, Hirai MY, Noji M, Saito K, Masuda T, Takamiya K, Shibata D, Ohta H (2005) Coordinated activation of metabolic pathways for antioxidants and defence compounds by jasmonates and their roles in stress tolerance in Arabidopsis. Plant J 44:653–668
49. Wolucka BA, Goossens A, Inze D (2005) Methyl jasmonate stimulates the de novo biosynthesis of vitamin C in plant cell suspensions. J Exp Bot 56:2527–2538
50. Shan CJ, Liang ZS (2010) Jasmonic acid regulates ascorbate and glutathione metabolism in *Agropyron cristatum* leaves under water stress. Plant Sci 178:130–139
51. Suza WP, Avila CA, Carruthers K, Kulkarni S, Goggin FL, Lorence A (2010) Exploring the impact of wounding and jasmonates on ascorbate metabolism. Plant Physiol Biochem 48:337–350
52. Pallanca JE, Smirnoff N (2000) The control of ascorbic acid synthesis and turnover in pea seedlings. J Exp Bot 51:669–674
53. Wolucka BA, Van Montagu M (2003) GDP-mannose 3′, 5′-epimerase forms GDP-L-gulose, a putative intermediate for the de novo biosynthesis of vitamin C in plants. J Biol Chem 278:47483–47490
54. Mieda T, Yabuta Y, Rapolu M, Motoki T, Takeda T, Yoshimura K, Ishikawa T, Shigeoka S (2004) Feedback inhibition of spinach L-galactose dehydrogenase by L-ascorbate. Plant Cell Physiol 45:1271–1279
55. Bonifacio A, Martins MO, Ribeiro CW, Fontenele AV, Carvalho FEL, Margis-Pinheiro M, Silveira JAG (2011) Role of peroxidases in the compensation of cytosolic ascorbate peroxidase knockdown in rice plants under abiotic stress. Plant Cell Environ 34:1705–1722
56. Rizhsky L, Hallak-Herr E, Van Breusegem F, Rachmilevitch S, Barr JE, Rodermel S, Inze D, Mittler R (2002) Double antisense plants lacking ascorbate peroxidase and catalase are less

sensitive to oxidative stress than single antisense plants lacking ascorbate peroxidase or catalase. Plant J 32:329–342

57. Zhang WY, Lorence A, Gruszewski HA, Chevone BI, Nessler CL (2009) AMR1, an Arabidopsis gene that coordinately and negatively regulates the Mannose/L-Galactose ascorbic acid biosynthetic pathway. Plant Physiol 150:942–950

Chapter 9
Ascorbate in Tomato, a Model Fruit

As one of the most important vegetable crops, the world-widely grown tomato provides rich sources of ascorbate for human's diet. The elucidation of genome sequence and short life cycle makes tomato a model fruit for fundamental research.

Since the D-Man/L-Gal pathway was proposed, several genes involved in this pathway, e.g. *GMP*, *GME*, *GGP*, *GPP*, *GalDH*, *GLDH*, *APX*, *AO*, *MDHAR*, *DHAR*, and *MIOX*, have been identified successively in tomato and utilized for metabolic engineering for ascorbate enchantment. In addition, Ascorbate oxidase promoter-binding protein (AOBP) was also cloned and functionally characterized in tomato. The present work on ascorbate in tomato has been focused on the gene isolation, expression analysis, and transgenic research. Like the ascorbate defective mutants in Arabidopsis, transgenic tomato lines with the genes related to ascorbate synthesis and metabolism show interesting growth phenotypes and altered stress sensitivity. However, the regulation mechanism for ascorbate synthesis and the potential alternative pathways such as Myoinositol pathway, Galacturonate pathway or Gulose pathway in tomato remain elusive.

The introgression line developed by crossing and backcrossing wild tomato species LA716 (*Solanum pennellii*) with the background M82 (*Solanum lycopersicum*) provide a potent tool for mapping and cloning genes involved in tomato ascorbate synthesis and metabolism [1]. Fourteen genes involved in ascorbate synthesis and metabolism in tomato were mapped in 9 chromosomes, respectively, and most of the genes were shown to be single copy in tomato genome [2]. The various tomato genetic populations facilitated the QTL mapping of ascorbate related genes in tomato. Through utilizing three different tomato populations, ascorbate related QTLs were simultaneously mapped on chromosomes 2, 8, 9, 10 and 12 [3]. Taken together the result of Zou et al. several genes of ascorbate biosynthesis coincided with QTLs. The recycling related gene *MDHAR* overlapped within QTLs on chromosome 9 (9-D), and biosynthesis related gene *GME* overlapped with QTLs on 9-J. More research on MDHAR indicated that the enzyme activity of MDHAR correlated closely with ascorbate accumulation in tomato under 4 °C, indicating a pivotal role of MDHAR in regulating ascorbate synthesis of tomato under stresses [4].

Y. Zhang, *Ascorbic Acid in Plants*, SpringerBriefs in Plant Science,
DOI: 10.1007/978-1-4614-4127-4_9, © The Author 2013

A comparable transcriptional analysis between a introgression line with high ascorbate accumulation, IL12-4, and its background parent, M82, indicated that a gene encoding pectinesterase and two genes encoding polygalacturonase were up-regulated in IL12-4, suggesting that more D-galacturonic acid was possibly produced by pectin degradation to promote ascorbate biosynthesis through galacturonic acid pathway [5].

The relevance between ascorbate accumulation and expression profiles of genes involved in ascorbate biosynthesis was investigated in different developmental stages of tomato fruits [6]. Among the 24 selected genes involved in ascorbate synthesis, oxidation and recycling, only *GPP* gene was shown to be expressed in close correlation with ascorbate accumulation, and *GPP* gene was also induced by post-harvest environmental stress, indicating *GPP* gene may play a pivotal role in ascorbate synthesis and post-harvest stress tolerance [6]. GGP and GPP catalyze two reaction steps, respectively, converting GDP-L-galactose to L-galactose, and the reaction product is specifically utilized for ascorbate synthesis. Transient overexpression of the kiwi *GGP* gene in tobacco resulted in >31-fold increase in ascorbate content of leaves [7].

Metabolic engineering of ascorbate biosynthesis was successively reported in tomato. The overexpression of *GMP* gene in tomato significantly improved the ascorbate content in tomato leaves and fruit, while *GMP* inhibition by RNAi suppression resulted in significantly lower ascorbate in transgenic tomato compared to wild-type control (unpublished data).

RNA interference of two *GME* genes was utilized to functionally characterize the *GME* in tomato. The *GME* suppressed transgenic tomato accumulated only 40–60 % ascorbate of the control in young leaves, and 60–80 % of the control in 20 day-after-pollination fruits, indicating the important role of *GME* in ascorbate synthesis in tomato. The *GME* suppressed tomato showed increased fragility and decreasing fruit firmness. The components of cell wall in leaves and fruits, particularly mannose and galactose related to the GME enzyme activity, were altered. The *GME* was probably involved in synthesis of ascorbate and noncellulosic cell wall polysaccharides [8]. Overexpressing *SlGME1* and *SlGME2* in tomato resulted in a significant increase in total ascorbate in leaves and red fruits compared with wild-type plants, with enhanced tolerance to oxidization, cold and salt stress [9].

The ascorbate content in the *GME2* co-suppression tomato plants was significantly reduced to only 12 and 28 % of the wild-type control in the leaves and fruit, respectively, showing the predominance of Smirnoff pathway for ascorbate synthesis in tomato [9]. Although the different plant species may prefer different pathway for ascorbate synthesis, e.g. the primary galacturonic acid pathway in strawberry fruit, so far, evidences from biochemical, genetic, and molecular biology showed that Smirnoff pathway is the major mechanism for ascorbate synthesis in most of the plant species.

The *SlGME2* co-suppression in transgenic tomato resulted in 1–2-fold increasing in gene expression of *SlPMI*, *SlGMP* and *SlGGP* in leaves, and 1.5–3.5-fold increasing in the gene expression of *SlGGP* in break stage fruit. In contrast, the overexpression of *GME1* and *GME2* in transgenic tomato led to obvious down-regulation of the gene expression of *SlPMI*, *SlGMP* and *SlGGP* in leaves, as well as down-regulated *SlGMP* in break stage fruits (unpublished data). These results suggest that there may be a regulatory mechanism in tomato to keep ascorbate accumulation at a state of homeostasis.

In addition, *SlGGP* showed the highest gene expression abundance in tomato leaves among all the genes involved in ascorbate synthesis. In break stage tomato fruit, *SlGGP* was also expressed at a high level following *SlGME1*. These results indicate *SlGGP* gene may play a key role in the regulation of tomato ascorbate synthesis. The VTC2 circle, composed of GME and GGP, was newly proposed for ascorbate synthesis [7]. The opposite expression abundance of *SlGMEs* and *SlGGP* in the *SlGMEs*-overexpressing tomato may be partly attributed to the VTC2 circle. GGP was reported to be localized in the nucleus as well as cytoplasm, suggesting GGP might play a regulatory role as a transcription factor.

GLDH gene suppression by RNAi approach did not alter the total ascorbate content in the leaves and fruits of tomato [10]. Feeding with substrate of GLDH enzyme, L-galactono-1,4-lactone, resulted in the same fold increasing of ascorbate in the *GLDH*-overexpressing transgenic tobacco and the untransformed control [11]. These results show that the GLDH-catalyzing step may not be a rate-limiting step. At the same time, many studies have shown that plant GalDH enzyme has high conversion efficiency for L-galactose, and exogenous L-galactose can be rapidly converted into ascorbate, indicating that the GalDH enzyme-catalyzing step may not be a rate-limiting step in ascorbate synthesis. GalDH and GLDH catalyze the last two reaction steps of Smirnoff pathway for ascorbate biosynthesis. The potential key steps for ascorbate synthesis in plants may be not GalDH or GLDH, but the upstream enzymes.

However, our unpublished data showed that overexpressing *GLDH* in tomato resulted in up to 3.6-fold increasing in GLDH activity compared to wild-type control. The *GLDH* overexpressing transgenic tomato showed 58 and 47 % increasing of ascorbate in leaves and fruits, respectively, compared to untransformed control (unpublished data). On the contrary, RNAi mediated GLDH suppression led to up to 54% decreasing of the GLDH activity, and 20 and 27 % decreasing of ascorbate content in leaves and fruits, respectively, compared to wild-type control (unpublished data). This is obviously inconsistent with previous observation by Alhagdow et al. and invites further investigation.

The RNAi mediated *APX* suppression in tomato resulted in 20–60 % decreasing of cytosolic APX activity compared to untransformed control. The down-regulated *APX* expression and APX enzyme activity resulted in 1.4- to 2.2-fold increase in ascorbate content in tomato fruits compared with control [12].

MDHAR and *DHAR* were cloned and over-expressed under a constitutive promoter in tomato, resulting in increased protein levels and enzymatic activity. Mature green and red ripe fruit with *DHAR* overexpressing showed 1.6-fold increase in ascorbate content in plants grown under relatively low light conditions. Conversely, *MDHAR* over-expressers had significantly reduced ascorbate levels in mature green fruits by 0.7 fold. Neither overexpressing line had altered levels of ascorbate in foliar tissues, indicating a complex regulation of the ascorbate pool size in tomato [13]. Our primary investigation showed, however, transgenic tomato overexpressing *MDHAR* had up to 3.4-fold increasing of MDHAR activity, and up to 21 % increasing in ascorbate content in leaves compared to wild-type control (unpublished data). Transgenic tomato overexpressing *DHAR* showed up to 5.2-fold increase in DHAR activity, and 58 and 57 % increasing in ascorbate levels in leaves and fruits, respectively, compared to wild-type control. However, *DHAR* suppression by RNAi in tomato did not cause alteration in DHAR activity or ascorbate level (unpublished data).

The galacturonic acid pathway of ascorbate synthesis was firstly proposed in the strawberry fruit. The ectopic expression of strawberry *GalUR* gene in tomato resulted in significantly improved ascorbate content in tomato, as well as enhanced resistance to salt stress in seedlings (unpublished data). Regulation of tomato ascorbate content can be achieved by overexpressing the strawberry *GalUR* gene. However, this result does not necessarily prove that galacturonic acid pathway of ascorbate synthesis is functional in tomato. Recently, gene expression profiles by chip hybridization on the introgression line IL12-4, with significantly and consistently higher ascorbate content than its parent, M82, did not give any difference in the expression abundance of genes involved in Smirnoff pathway, but showed that the transcriptional abundance of genes encoding a pectinesterase and two polygalacturonase in IL12-4 were significantly higher than that in M82 [5]. The authors speculated that increased gene expression accelerates the degradation of pectin and the formation of D-galacturonic acid. The increasing substrate content will enhance the synthesis efficiency of ascorbate by galacturonic acid pathway, resulting in higher ascorbate level in the introgression line than its parent. However, where this hypothesis of D-galacturonic acid pathway is valid only in IL12-4, or universal in tomato species remains unclear. In order to confirm the D-galacturonic acid synthesis pathway of ascorbate is prevalent in tomato, the tomato homologous gene of strawberry *GalUR* ought to be cloned and functionally characterized by overexpression and gene suppression in tomato. With the completion of tomato genome sequencing, we found a candidate gene for tomato *GalUR*, SGN-U580238, by blasting tomato Unigene database with the amino acid sequence of *GalUR* in strawberry and kiwi. The functional investigation the gene is underway. Also, feeding plant with precursors may help to determine the potential D-galacturonic acid pathway for ascorbate biosynthesis in tomato.

In addition, several other genes involved tomato ascorbate biosynthesis, such as *GGP*, *GPP*, *AO*, and *MIOX*, are cloned and unitized for ascorbate metabolic engineering (unpublished data).

The transgenic results above show several genes in Smirnoff pathway of tomato ascorbate synthesis have potent roles in modulating ascorbate synthesis, while some of them are not effectively regulated. Simultaneous regulation of the genes in the ascorbate synthesis pathway may provide a more effective approach to improve ascorbate content in tomato. At same time, the transcription factors or regulators regulating the ascorbate synthesis are emerging as research focus, e.g. AMR1 [14], as they can regulate more than one gene and produce more efficiency than the enzyme encoding genes.

References

1. Eshed Y, Zamir D (1995) An introgression line population of *Lycopersicon pennellii* in the cultivated tomato enables the identification and fine mapping of yield-associated QTL. Genetics 141:1147–1162
2. Zou LP, Li HX, Ouyang B, Zhang JH, Ye ZB (2006) Cloning and mapping of genes involved in tomato ascorbic acid biosynthesis and metabolism. Plant Sci 170:120–127
3. Stevens R, Buret M, Duffe P, Garchery C, Baldet P, Rothan C, Causse M (2007) Candidate genes and quantitative trait loci affecting fruit ascorbic acid content in three tomato populations. Plant Physiol 143:1943–1953
4. Stevens R, Page D, Gouble B, Garchery C, Zamir D, Causse M (2008) Tomato fruit ascorbic acid content is linked with monodehydroascorbate reductase activity and tolerance to chilling stress. Plant Cell Environ 31:1086–1096
5. Di Matteo A, Sacco A, Anacleria M, Pezzotti M, Delledonne M, Ferrarini A, Frusciante L, Barone A (2010) The ascorbic acid content of tomato fruits is associated with the expression of genes involved in pectin degradation. BMC Plant Biology 10:163
6. Ioannidi E, Kalamaki MS, Engineer C, Pateraki I, Alexandrou D, Mellidou I, Giovannonni J, Kanellis AK (2009) Expression profiling of ascorbic acid-related genes during tomato fruit development and ripening and in response to stress conditions. J Exp Bot 60:663–678
7. Laing WA, Wright MA, Cooney J, Bulley SM (2007) The missing step of the L-galactose pathway of ascorbate biosynthesis in plants, an L-galactose guanyltransferase, increases leaf ascorbate content. Proc Nat Acad Sci USA 104:9534–9539
8. Gilbert L, Alhagdow M, Nunes-Nesi A, Quemener B, Guillon F, Bouchet B, Faurobert M, Gouble B, Page D, Garcia V, Petit J, Stevens R, Causse M, Fernie AR, Lahaye M, Rothan C, Baldet P (2009) GDP-D-mannose 3,5-epimerase (GME) plays a key role at the intersection of ascorbate and non-cellulosic cell-wall biosynthesis in tomato. Plant J 60:499–508
9. Zhang CJ, Liu JX, Zhang YY, Cai XF, Gong PJ, Zhang JH, Wang TT, Li HX, Ye ZB (2011a) Overexpression of *SlGMEs* leads to ascorbate accumulation with enhanced oxidative stress, cold, and salt tolerance in tomato. Plant Cell Rep 30:389–398
10. Alhagdow M, Mounet F, Gilbert L, Nunes-Nesi A, Garcia V, Just D, Petit J, Beauvoit B, Fernie AR, Rothan C, Baldet P (2007) Silencing of the mitochondrial ascorbate synthesizing enzyme L-galactono-1,4-lactone dehydrogenase affects plant and fruit development in tomato. Plant Physiol 145:1408–1422
11. Imai T, Niwa M, Ban Y, Hirai M, Oba K, Moriguchi T (2009) Importance of the L-galactonolactone pool for enhancing the ascorbate content revealed by L-galactonolactone dehydrogenase-overexpressing tobacco plants. Plant Cell Tiss Org Cult 96:105–112
12. Zhang YY, Li HX, Shu WB, Zhang CJ, Ye ZB (2011b) RNA interference of a mitochondrial *APX* gene improves vitamin C accumulation in tomato fruit. Sci Hort 129:220–226
13. Haroldsen VM, Chi-Ham CL, Kulkarni S, Lorence A, Bennett AB (2011) Constitutively expressed DHAR and MDHAR influence fruit, but not foliar ascorbate levels in tomato. Plant Physiol Biochem 49:1244–1249
14. Zhang WY, Lorence A, Gruszewski HA, Chevone BI, Nessler CL (2009) AMR1, an Arabidopsis gene that coordinately and negatively regulates the Mannose/L-Galactose ascorbic acid biosynthetic pathway. Plant Physiol 150:942–950

Chapter 10
Metabolic Modification of Ascorbate in Plants

Ascorbate is the major soluble antioxidant found in plants and is also an essential component of human nutrition. Evidence suggests that the plasma levels of ascorbate in large sections of the population are sub-optimal for the health protective effects of this vitamin [1].

Elucidation of the ascorbate biosynthetic pathway now opens the way to manipulating ascorbate biosynthesis in plants. Theoretically, any factors influencing ascorbate biosynthesis and metabolism can ultimately affect in vivo accumulation of ascorbate in plants, and the genes and corresponding enzymes involved in ascorbate biosynthesis and metabolism can be regulated to achieve the optimal accumulation of ascorbate [2].

Our increasing understanding of ascorbate synthesis and accumulation in plants will facilitate engineering plant ascorbate. Several possible strategies could be followed to increase ascorbate production, such as enhancing the pivotal steps or overcoming the rate limiting steps in the biosynthetic pathway, promoting recycling, and reducing catabolism. Therefore, the purpose of improving ascorbate content in plants can probably be achieved by overexpressing genes involved in ascorbate biosynthesis and recycling, such as *GME* for synthesis, and *DHAR* and *MDHAR* for recycling. On the other hand, inhibition of the gene expression and enzyme activity responsible for ascorbate oxidation and catabolism, such as *APX* and *AO*, may also achieve the goal of improving ascorbate. The emerging transcription factors or regulators for ascorbate accumulation may hold more promise.

The ascorbate metabolic engineering has been very successful from yeast [3] to plants. The existing and potential achievements in increasing ascorbate production would provide the opportunity for enhancing nutritional quality and stress tolerance of crop plants.

Y. Zhang, *Ascorbic Acid in Plants*, SpringerBriefs in Plant Science, DOI: 10.1007/978-1-4614-4127-4_10, © The Author 2013

10.1 Overexpression and Ectopic Expression

Genetic engineering approaches have been widely adopted to modify the ascorbate biosynthesis and metabolism, and enhance ascorbate accumulation in various plant species.

Transient expression of *PMM* in tobacco using viral vector-mediated ectopic expression led to a 20–50 % increase in ascorbate content. And transgenic expression of *PMM* in Arabidopsis also increased ascorbate content by 25–33 % [4]. Transgenic tobacco plants overexpressing the acerola *PMM* showed about 2-fold increase in ascorbate contents compared with the wild-type, with a corresponding increase in the *PMM* transcript levels and enzyme activities [5].

The gene members of *GME* family, such as *GME1* and *GME2* from *Solanum lycopersicum* and *GME2* from *Solanum pennellii*, were utilized for overexpressing in tomato. *SlGME1* overexpressing resulted in 56 % and 61 % increasing of ascorbate content in leaves and fruits, respectively in tomato. Transgenic tomato with overexpression of *SlGME2* showed 31 % and 57 % increasing in ascorbate content in leaves and fruits respectively. Overexpressing *SpGME2* in tomato resulted in 54 % and 103 % increasing of ascorbate accumulation in leaves and fruits respectively [6].

Transient expressing a gene encoding GDP-L-galactose-hexose-1-phosphate guanyltransferase (catalyzing the transfer of GMP from GDP-L-galactose to a hexose-1-phosphate) from kiwifruit results in >3-fold increase of ascorbate in tobacco leaves as well as a 50-fold increase in the corresponding enzyme activity [7]. Overexpression of the kiwifruit GDP-L-galactose guanyltransferase gene in Arabidopsis resulted in up to 4-fold increase in ascorbate, while up to 7-fold increase in ascorbate was observed when both GDP-L-galactose guanyltransferase and GDP-mannose-3′,5′-epimerase genes were co-expressed [8].

GLDH catalyzes the final step of ascorbate synthesis in plants. *GLDH* gene is found be single copy in the genome of melon [9] and tomato [10], which will facilitate the gene regulation in plants. Overexpressing *GLDH* gene in transgenic tomato resulted increasing GLDH activity and ascorbate content in both leaves and fruits compared to the untransformed control, while the RNAi suppression of *GLDH* led to the lowered GLDH enzyme activity and ascorbate content in leaves and fruits. Activity alteration of other enzymes involved in ascorbate biosynthesis was not observed in the transgenic tomato plants with *GLDH* regulation, indicating the changing ascorbate level resulted mainly from of the *GLDH* regulation (unpublished data). The similar *GLDH* regulation effect was also observed in transgenic rice [11]. Regulating the gene expression of *GLDH* and its enzyme activity is an effective approach to improve the ascorbate accumulation in plants.

Overexpressing the gene encoding L-gulono-1,4-lactone oxidase, the final step enzyme for ascorbate synthesis from animal, resulted in improved ascorbate content in transgenic lettuce and tobacco [12]. The plants are supposed to share the ascorbate biosynthesis mechanism of the gulose pathway in animals. L-galactono-1,4-lactone and L-gulono-1,4-lactone are both direct precursors for ascorbate synthesis, under the catalyzing of GLDH and GLOase, respectively (Fig. 3.1).

Overexpression of *MIOX* led to 2–3-fold increasing of ascorbate in transgenic Arabidopsis [13]. In addition, the overexpression of strawberry *MIOX* gene in tomato resulted in 38–61 % improvement of ascorbate content compared to the untransformed control when fed with myoinositol (unpublished data). This also suggested the potential myoinositol pathway for ascorbate synthesis in tomato.

Overexpressing the gene encoding D-galacturonic acid reductase from strawberry resulted in 2–3-fold increasing of ascorbate in Arabidopsis [14]. Ectopic expression of strawberry *GalUR* gene in potato resulted in 1.6–2-fold increase in ascorbate content accompanied by the increasing GalUR enzyme activity [15]. HPLC analysis of transgenic tomato overexpressing *FaGalUR* from strawberry showed that some of the *FaGalUR* overexpressing lines had higher content (28.7–65.2 %) of ascorbate than wild-type control. Correlation between the ascorbate accumulation and the *FaGalUR* expression level was observed (unpublished data).

Once used, the ascorbate can be regenerated from its oxidized form in a reaction catalyzed by DHAR. DHAR is a key enzyme promoting ascorbate regeneration in plants. Overexpression of *DHAR* in plants can also increase the level of ascorbate through improved ascorbate recycling. Overexpressing the *DHAR1* gene in tomato resulted in up to 5.2-fold increasing in DHAR activity, and 58 %- and 57 %-increasing in ascorbate accumulation of leaves and fruits, respectively. The ascorbate increasing in leaves and fruits was mainly caused by DHAR regulation as other enzyme activities for ascorbate synthesis were not altered significantly (unpublished data). Overexpressing the wheat *DHAR* gene in tobacco and maize resulted in 32- and 100-fold increasing in DHAR activity, respectively, and 2–4-fold increasing in the ascorbate content in leaves and maize grain [16]. In addition, the ascorbate redox state in both *DHAR*-overexpressing tobacco and maize was significantly increased. The level of glutathione, the reductant used by DHAR, was also increased, as did its redox state [16]. The overexpression of cytosolic DHAR significantly increased DHAR activities and ascorbate contents in potato leaves and tubers, whereas chloroplastic *DHAR* overexpression only increased DHAR activities and ascorbate contents in leaves, and did not change them in tubers [17].

The MDHAR is another important enzyme promoting ascorbate regeneration. Transgenic tobacco plants overexpressing the Acerola *MDHAR* gene showed increased DHAR activity as well as improved ascorbate accumulation [18]. Overexpressing *MDHAR1* gene in tomato showed a similar regulatory effect on the ascorbate accumulation with *DHAR1* overexpression. The ascorbate in the leaves of *MDHAR*-overexpressing tomato plants was higher than that of wild-type control. However, one of the transgenic tomato lines with sense expression of *MDHAR* showed decreasing MDHAR activity and ascorbate content, probably because of the gene co-suppression (unpublished data). These results demonstrate the feasibility of modulating ascorbate accumulation in plants through regulating ascorbate recycling.

All these results show that it is feasible and effective to modulate the ascorbate accumulation in plants by regulating the expression of genes responsible for ascorbate synthesis. Overexpression strategy is suitable for regulating genes involved in ascorbate biosynthesis to improve ascorbate content in plants.

10.2 Gene Suppression

After being synthesized in plants, ascorbate is subject to catabolism and oxidation. The enzymes responsible for ascorbate catabolism can be inhibited to prevent the ascorbate degradation and thus increase its final accumulation in plants. Only a few of the enzymes for ascorbate catabolism are identified, including APX and AO.

RNAi suppression of tomato APX gene resulted in 20–60 % decreasing in cytosolic APX activity, and 55–71 % decreasing in mitochondrial APX enzyme activity compared to wild-type plants. The APX suppressed tomato showed 1.4- to 2.2-fold increase in ascorbate content in tomato fruits compared to wild-type control [19]. The decreased AO enzyme activity and significantly improved ascorbate content in tomato fruit were observed in transgenic tomato with AO suppression by RNAi [20]. Antisense expression of AO gene in transgenic tobacco plants resulted in 2.5-fold decreasing in AO activity, and 42 % increasing in reduced state of ascorbic acid in apoplast compared to the control [21]. This demonstrates the feasibility to improve ascorbate content in plants by suppressing genes related to ascorbate catabolism.

However, suppressing the enzymes responsible for ascorbate synthesis will lead to decreasing ascorbate level. Antisense inhibition of potato GMP gene led to lowered ascorbate content in transgenic plants [22]. Antisense expression of gene encoding L-galactose dehydrogenase resulted in reduced ascorbate accumulation in transgenic Arabidopsis [23]. Down-regulated expression of gene encoding phosphomannose isomerase one through RNA interference resulted in a substantial decrease in the total ascorbate content in Arabidopsis leaves [24]. It should be noted that RNAi suppression of DHAR did not affect the ascorbate accumulation in transgenic tomato, probably because of multi-gene family for DHAR isozyme or chromosomal location of the T-DNA (unpublished data).

Although the catalyzing steps in D-Man/L-Gal pathway have been elucidated and several alternative pathways have been proposed, it is supposed that regulation of ascorbate biosynthesis might occur at more than one step and warrants further investigation to allow for the efficient manipulation of ascorbate levels in plants [25]. In addition, because ascorbate is not stable and easily degraded during processing or storage, future focus might be given to minimizing ascorbate losses prior to ingestion as well as improving the ascorbate accumulation in plants.

References

1. Davey MW, Van Montagu M, Inze D, Sanmartin M, Kanellis A, Smirnoff N, Benzie IJJ, Strain JJ, Favell D, Fletcher J (2000) Plant L-ascorbic acid: chemistry, function, metabolism, bioavailability and effects of processing. J Sci Food Agric 80:825–860
2. Ishikawa T, Dowdle J, Smirnoff N (2006) Progress in manipulating ascorbic acid biosynthesis and accumulation in plants. Physiologia Plantarum 126: 343–355
3. Fossati T, Solinas N, Porro D, Branduardi P (2011) L-ascorbic acid producing yeasts learn from plants how to recycle it. Metab Eng 13:177–185
4. Qian WQ, Yu CM, Qin HJ, Liu X, Zhang AM, Johansen IE, Wang DW (2007) Molecular and functional analysis of phosphomannomutase (PMM) from higher plants and genetic evidence for the involvement of PMM in ascorbic acid biosynthesis in Arabidopsis and *Nicotiana benthamiana*. Plant J 49:399–413
5. Badejo AA, Eltelib HA, Fukunaga K, Fujikawa Y, Esaka M (2009) Increase in ascorbate content of transgenic tobacco plants overexpressing the acerola (*Malpighia glabra*) phosphomannomutase gene. Plant Cell Physiol 50:423–428
6. Zhang CJ, Liu JX, Zhang YY, Cai XF, Gong PJ, Zhang JH, Wang TT, Li HX, Ye ZB (2011a) Overexpression of *SlGMEs* leads to ascorbate accumulation with enhanced oxidative stress, cold, and salt tolerance in tomato. Plant Cell Rep 30:389–398
7. Laing WA, Wright MA, Cooney J, Bulley SM (2007) The missing step of the L-galactose pathway of ascorbate biosynthesis in plants, an L-galactose guanyltransferase, increases leaf ascorbate content. Proc Nat Acad Sci U S A 104:9534–9539
8. Bulley SM, Rassam M, Hoser D, Otto W, Schunemann N, Wright M, MacRae E, Gleave A, Laing W (2009) Gene expression studies in kiwifruit and gene over-expression in Arabidopsis indicates that GDP-L-galactose guanyltransferase is a major control point of vitamin C biosynthesis. J Exp Bot 60:765–778
9. Pateraki I, Sanmartin M, Kalamaki MS, Gerasopoulos B, Kanellis AK (2004) Molecular characterization and expression studies during melon fruit development and ripening of L-galactono-1,4-lactone dehydrogenase. J Exp Bot 55:1623–1633
10. Zou LP, Li HX, Ouyang B, Zhang JH, Ye ZB (2006) Cloning and mapping of genes involved in tomato ascorbic acid biosynthesis and metabolism. Plant Sci 170:120–127
11. Liu YH, Yu L, Wang RZ (2011) Level of ascorbic acid in transgenic rice for L-galactono-1, 4-lactone dehydrogenase overexpressing or suppressed is associated with plant growth and seed set. Acta Physiol Plant 33:1353–1363
12. Jain AK, Nessler CL (2000) Metabolic engineering of an alternative pathway for ascorbic acid biosynthesis in plants. Mol Breed 6:73–78
13. Lorence A, Chevone BI, Mendes P, Nessler CL (2004) Myo-inositol oxygenase offers a possible entry point into plant ascorbate biosynthesis. Plant Physiol 134:1200–1205
14. Agius F, Gonzalez-Lamothe R, Caballero JL, Munoz-Blanco J, Botella MA, Valpuesta V (2003) Engineering increased vitamin C levels in plants by overexpression of a D-galacturonic acid reductase. Nat Biotechnol 21:177–181
15. Hemavathi, Upadhyaya CP, Young KE, Akula N, Kim HS, Heung JJ, Oh OM, Aswath CR, Chun SC, Kim DH, Park SW (2009) Over-expression of strawberry D-galacturonic acid reductase in potato leads to accumulation of vitamin C with enhanced abiotic stress tolerance. Plant Sci 177: 659–667
16. Chen Z, Young TE, Ling J, Chang SC, Gallie DR (2003) Increasing vitamin C content of plants through enhanced ascorbate recycling. Proc Nat Acad Sci U S A 100:3525–3530
17. Qin AG, Shi QH, Yu XC (2011) Ascorbic acid contents in transgenic potato plants overexpressing two dehydroascorbate reductase genes. Mol Biol Rep 38:1557–1566
18. Eltelib HA, Fujikawa Y, Esaka M (2012) Overexpression of the acerola (*Malpighia glabra*) monodehydroascorbate reductase gene in transgenic tobacco plants results in increased ascorbate levels and enhanced tolerance to salt stress. S Afr J Bot 78:295–301

19. Zhang YY, Li HX, Shu WB, Zhang CJ, Ye ZB (2011b) RNA interference of a mitochondrial
    APX gene improves vitamin C accumulation in tomato fruit. Sci Hort 129:220–226
20. Zhang YY, Li HX, Shu WB, Zhang CJ, Zhang W, Ye ZB (2011c) Suppressed expression of
    ascorbate oxidase gene promotes ascorbic acid accumulation in tomato fruit. Plant Mol Biol
    Rep 29:638–645
21. Pignocchi C, Foyer CH (2003) Apoplastic ascorbate metabolism and its role in the regulation
    of cell signalling. Curr Opin Plant Biol 6:379–389
22. Keller R, Springer F, Renz A, Kossmann J (1999) Antisense inhibition of the GDP-mannose
    pyrophosphorylase reduces the ascorbate content in transgenic plants leading to
    developmental changes during senescence. Plant J 19:131–141
23. Gatzek S (2002) Antisense suppression of L-galactose dehydrogenase in *Arabidopsis thaliana*
    provides evidence for its role in ascorbate-synthesis and reveals light modulated L-galactose
    synthesis. Plant J 31:541–553
24. Maruta T, Yonemitsu M, Yabuta Y, Tamoi M, Ishikawa T, Shigeoka S (2008) Arabidopsis
    phosphomannose isomerase 1, but not phosphomannose isomerase 2, is essential for ascorbic
    acid biosynthesis. J Biol Chem 283:28842–28851
25. Linster CL, Clarke SG (2008) L-Ascorbate biosynthesis in higher plants: the role of VTC2.
    Trends Plant Sci 13:567–573

Chapter 11
Regulating Ascorbate Biosynthesis and Metabolism for Stress Tolerance in Plants

Drought, salt, extreme temperature, and other abiotic stress causes severe damage to crop production. Due to genetic complexity of stress resistance and the lack of an effective selection, little progress has been made in conventional resistance breeding of crops. The deepened insights into plant stress resistance mechanism and development of molecular breeding techniques make it possible to cultivate crops with stress tolerance by genetic engineering.

Excessive ROS are usually generated in plants under the adverse conditions. The mechanism of scavenging ROS in plants mainly largely lies in the various enzymatic and non-enzymatic antioxidants. Ascorbate and its metabolism related enzymes play an important role in scavenging ROS. Ascorbate is directly involved in the clearance of ROS as an electron donor, and hydrogen peroxide can also be cleared by APX. Due to the lack of CAT in chloroplasts, APX is crucial for the hydrogen peroxide removal in the chloroplast. Other enzymes, such as enzymes in ascorbate–glutathione cycle involved in ascorbate metabolism, contribute greatly to plant capacity of removing ROS.

Generally, overexpression of genes responsible for ascorbate synthesis or regeneration, such as *GMP*, *GME* and *GalUR*, or *DHAR* and *MDHAR* respectively, will likely improve plants ascorbate content as well as stress resistance. On the contrary, overexpressing genes responsible for ascorbate catabolism, such as *APX*, will potentially improve plant stress resistance, but reduce the ascorbate content in plants.

Improving the content of ascorbate in plants can help enhance the plant resistance to abiotic stress. Seedling of transgenic tomato overexpressing *SlGME1* or *SlGME2* showed improved tolerance to salt, drought and oxidative stress compared to the untransformed control, indicating that overexpression of *GME* gene of ascorbate synthesis pathway can improve ascorbate accumulation as well as the stress resistance [1]. The gene expression profiles of *SlGME1* overexpressing tomato showed that several peroxidase encoding genes were up-regulated in break stage fruits, suggesting that the antioxidant system in *SlGME1* overexpressing tomato has been activated [1]. The transgenic plants overexpressing *SlGMP* also showed improved

Fig. 11.1 Germination rate of transgenic tomato overexpressing *DHAR* (*right*) and wild-type control (*left*) on MS medium with 100 mmol/L NaCl (courtesy of Professor Zhibiao Ye and Dr. Liping Zou)

tolerance to oxidative stress. However, the *SlGME2* co-suppression in tomato led to reduced accumulation of ascorbate and extreme sensitivity to oxidative stress (unpublished data).

Overexpression of *DHAR* in transgenic tobacco plants showed increasing resistance to ozone [2]. Overexpression *DHAR1* gene in tomato resulted in improved tolerance to salt and the herbicide paraquat to some extent (unpublished data). Seeds of transgenic tomato overexpressing *DHAR1* as well as untransformed control were germinated on the MS medium containing 100 mmol/L NaCl, and the germination rate 1 week after sowing of the *DHAR1* overexpressing plants and the untransformed control was 70 and 20 %, respectively (Fig. 11.1), indicating that the *DHAR* overexpressing helps to improve salt tolerance in plants.

Paraquat is an herbicide producing oxidative stress to plants. Overexpression of *DHAR1* in transgenic tomato showed improved paraquat resistance compared to the untransformed control. The leaf discs of transgenic tomato overexpressing *DHAR1* as well as the wild-type control were treated with 2 μmol/L paraquat for 24 h, and the leaf chlorosis of the untransformed control was more severe than that of the *DHAR1* overexpressing line (Fig. 11.2), suggesting that *DHAR1*-overexpressing transgenic plants were conferred improved resistance to oxidative stress caused by paraquat.

Overexpressing the acerola *MDHAR* gene in tobacco led to greater amounts of ascorbate and higher MDHAR activity than in the control plants. Lipid peroxidation and chlorophyll degradation, which were stimulated in control plants, were restrained in *MDHAR* transgenic plants subjected to salt stress. These results indicate that overexpression of acerola *MDHAR* imparts greater tolerance to salt stress [3].

Overexpressing genes involved the alternative pathway for ascorbate synthesis is also shown to enhance stress tolerance in plants. Over-expression of strawberry D-galacturonic acid reductase in potato lead to accumulation of ascorbate with enhanced tolerance to abiotic stresses induced by methyl viologen, NaCl or mannitol as compared to untransformed control plants [4]. Further biochemical investigation showed that in the transgenic potato overexpressing the strawberry *GalUR*, increased levels of antioxidant enzymes such as SOD, CAT, APX, DHAR, and GR were observed, and the content of ascorbate, glutathione and proline was improved compared to untransformed control upon various abiotic stress

Fig. 11.2 Chlorophyll bleaching of the leaf discs of transgenic tomato overexpressing *DHAR* (*right*) and wild-type control (*left*) under paraquat treatment (courtesy of Professor Zhibiao Ye and Dr. Liping Zou)

treatment. The redox state of ascorbate and glutathione was also altered in the *GalUR* overexpressing potato. These evidence indicate that improved ascorbate level activates the antioxidative system to defend against abiotic stresses [5]. Ectopic expression of strawberry *GalUR* in tomato also resulted in enhanced salt tolerance (unpublished data).

Additionally, ectopic expression of rat L-gulono-1, 4-lactone oxidase gene in potato resulted in higher survival rate after a variety of abiotic stress treatment compared to the control [6]. The overexpression of *MIOX* in transgenic tomato has resulted in the increasing antioxidative capacity as well as improved ascorbate content (unpublished data).

Numerous studies have shown that the regulation of *APX* gene expression can improve the plant resistance to stresses [7–16, 18, 19]. Overexpressing Arabidopsis *APX3* gene in tobacco resulted in enhanced CO_2 assimilation and seed production under water deficit [10]. Overproduction of pepper APX in tobacco plants increased peroxidase activity that enhances ROS scavenging capacity, leading to oxidative stress tolerance and oomycete pathogen resistance [12]. Enhanced tolerance against oxidative stress and high temperature was observed in transgenic potato plants expressing both *APX* and *SOD* in chloroplasts [13]. Transgenic Arabidopsis plants constitutively overexpressing a barley peroxisomal *APX* gene showed enhanced zinc and cadmium tolerance and accumulation compared to wild-type control [15]. Over-expression of a *Populus* peroxisomal *APX* gene in tobacco plants resulted in enhanced stress tolerance [16]. Expression of Cu/Zn SOD and APX in chloroplasts of sweet potato enhanced drought resistance and capacity for recovery from drought stress [17]. Simultaneous expression of choline oxidase (ChOx), SOD and APX in potato chloroplasts provides synergistically enhanced protection against oxidation, salt and drought stresses [18]. *Brassica napus* plants overexpressing *tAPX* showed improved resistance to salt and drought stress in plants [20]. Overexpression of the ascorbate peroxidase gene conferred enhanced chilling tolerance at the booting stage in rice. Higher APX activity enhances hydrogen peroxide-scavenging capacity and protects spikelets from lipid peroxidation, thereby increasing spikelet fertility under cold stress [21]. Overexpressing the stromal *APX* from the euhalophyte *Suaeda salsa* in Arabidopsis resulted in longer roots, higher total chlorophyll content, less cell membrane damage, and lower hydrogen peroxide level under NaCl stress as compared to wild-type control [19].

In addition, inhibition of apoplastic *AO* gene expression can increase the salt tolerance of antisense transgenic tobacco and T-DNA inserted Arabidopsis mutant [22].

In short, regulating the expression of genes involved in ascorbate biosynthesis and metabolism can modulate ascorbate level and enzymes activities related to ascorbate metabolism, improve ROS scavenging capacity, and thereby enhance the plant resistance to stress. Therefore, the regulation of ascorbate biosynthesis and metabolism is another feasible way for engineering stress tolerance in plants.

References

1. Zhang CJ, Liu JX, Zhang YY, Cai XF, Gong PJ, Zhang JH, Wang TT, Li HX, Ye ZB (2011) Overexpression of *SlGMEs* leads to ascorbate accumulation with enhanced oxidative stress, cold, and salt tolerance in tomato. Plant Cell Rep 30:389–398
2. Chen Z, Gallie DR (2005) Increasing tolerance to ozone by elevating foliar ascorbic acid confers greater protection against ozone than increasing avoidance. Plant Physiol 138:1673–1689
3. Eltelib HA, Fujikawa Y, Esaka M (2012) Overexpression of the acerola (*Malpighia glabra*) monodehydroascorbate reductase gene in transgenic tobacco plants results in increased ascorbate levels and enhanced tolerance to salt stress. S Afr J Bot 78:295–301
4. Hemavathi, Upadhyaya CP, Young KE, Akula N, Kim HS, Heung JJ, Oh OM, Aswath CR, Chun SC, Kim DH, Park SW (2009) Over-expression of strawberry D-galacturonic acid reductase in potato leads to accumulation of vitamin C with enhanced abiotic stress tolerance. Plant Sci 177:659–667
5. Hemavathi, Upadhyaya CP, Akula N, Kim HS, Jeon JH, Ho OM, Chun SC, Kim DH, Park SW (2011) Biochemical analysis of enhanced tolerance in transgenic potato plants overexpressing D-galacturonic acid reductase gene in response to various abiotic stresses. Mol Breed 28:105–115
6. Hemavathi, Upadhyaya CP, Akula N, Young KE, Chun SC, Kim DH, Park SW (2010) Enhanced ascorbic acid accumulation in transgenic potato confers tolerance to various abiotic stresses. Biotechnol Lett 32:321–330
7. Kwon SY, Jeong YJ, Lee HS, Kim JS, Cho KY, Allen RD, Kwak SS (2002) Enhanced tolerances of transgenic tobacco plants expressing both superoxide dismutase and ascorbate peroxidase in chloroplasts against methyl viologen-mediated oxidative stress. Plant Cell Environ 25:873–882
8. Danna CH, Bartoli CG, Sacco F, Ingala LR, Santa-Maria GE, Guiamet JJ, Ugalde RA (2003) Thylakoid-bound ascorbate peroxidase mutant exhibits impaired electron transport and photosynthetic activity. Plant Physiol 132:2116–2125
9. Pnueli L, Liang H, Rozenberg M, Mittler R (2003) Growth suppression, altered stomatal responses, and augmented induction of heat shock proteins in cytosolic ascorbate peroxidase (APX1)-deficient Arabidopsis plants. Plant J 34:185–201
10. Yan JQ, Wang J, Tissue D, Holaday AS, Allen R, Zhang H (2003) Photosynthesis and seed production under water-deficit conditions in transgenic tobacco plants that overexpress an Arabidopsis ascorbate peroxidase gene. Crop Sci 43:1477–1483
11. Murgia I, Tarantino D, Vannini C, Bracale M, Carravieri S, Soave C (2004) *Arabidopsis thaliana* plants overexpressing thylakoidal ascorbate peroxidase show increased resistance to Paraquat-induced photooxidative stress and to nitric oxide-induced cell death. Plant J 38:940–953

12. Sarowar S, Kim EN, Kim YJ, Ok SH, Kim KD, Hwang BK, Shin JS (2005) Overexpression of a pepper ascorbate peroxidase-like 1 gene in tobacco plants enhances tolerance to oxidative stress and pathogens. Plant Sci 169:55–63

13. Tang L, Kwon SY, Kim SH, Kim JS, Choi JS, Cho KY, Sung CK, Kwak SS, Lee HS (2006) Enhanced tolerance of transgenic potato plants expressing both superoxide dismutase and ascorbate peroxidase in chloroplasts against oxidative stress and high temperature. Plant Cell Rep 25:1380–1386

14. Lee SH, Ahsan N, Lee KW, Kim DH, Lee DG, Kwak SS, Kwon SY, Kim TH, Lee BH (2007) Simultaneous overexpression of both Cu Zn superoxide dismutase and ascorbate peroxidase in transgenic tall fescue plants confers increased tolerance to a wide range of abiotic stresses. J Plant Physiol 164:1626–1638

15. Xu WF, Shi WM, Liu F, Ueda A, Takabe T (2008) Enhanced zinc and cadmium tolerance and accumulation in transgenic Arabidopsis plants constitutively overexpressing a barley gene (*HvAPX1*) that encodes a peroxisomal ascorbate peroxidase. Botany 86:567–575

16. Li YJ, Hai RL, Du XH, Jiang XN, Lu H (2009) Over-expression of a *Populus* peroxisomal ascorbate peroxidase (PpAPX) gene in tobacco plants enhances stress tolerance. Plant Breed 128:404–410

17. Lu YY, Deng XP, Kwak SS (2010) Over expression of CuZn superoxide dismutase (CuZn SOD) and ascorbate peroxidase (APX) in transgenic sweet potato enhances tolerance and recovery from drought stress. Afr J Biotechnol 9:8378–8391

18. Ahmad R, Kim YH, Kim MD, Kwon SY, Cho K, Lee HS, Kwak SS (2010) Simultaneous expression of choline oxidase, superoxide dismutase and ascorbate peroxidase in potato plant chloroplasts provides synergistically enhanced protection against various abiotic stresses. Physiol Plantarum 138:520–533

19. Li K, Pang CH, Ding F, Sui N, Feng ZT, Wang BS (2012) Overexpression of *Suaeda salsa* stroma ascorbate peroxidase in Arabidopsis chloroplasts enhances salt tolerance of plants. S Afr J Botany 78:235–245

20. Wang JM, Fan ZY, Liu ZB, Xiang JB, Chai L, Li XF, Yang Y (2011) Thylakoid-bound ascorbate peroxidase increases resistance to salt stress and drought in *Brassica napus*. Afr J Biotechnol 10:8039–8045

21. Sato Y, Masuta Y, Saito K, Murayama S, Ozawa K (2011) Enhanced chilling tolerance at the booting stage in rice by transgenic overexpression of the ascorbate peroxidase gene, *OsAPXa*. Plant Cell Rep 30:399–406

22. Yamamoto A, Bhuiyan NH, Waditee R, Tanaka Y, Esaka M, Oba K, Jagendorf AT, Takabe T (2005) Suppressed expression of the apoplastic ascorbate oxidase gene increases salt tolerance in tobacco and Arabidopsis plants. J Exp Bot 56:1785–1796